THE
WAR
IN THE
GARDEN OF EDEN

"Madman! Look through my eyes if thou hast none of thine own."

This Book Belongs to:

SA PIEN TIA

SCI EN TIA

IN TELLEC TVS

FOR TITV DO

CON SILI VM

PI E TAS

TIMOR DÑI

THE
WAR
IN THE
GARDEN OF EDEN

BY

KERMIT ROOSEVELT

*Sketch portrait of Kermit Roosevelt
by John S. Sargent*

TABLE OF CONTENTS

Publisher's Note

Blessed beloved bookreader, you have found this volume in your vision. We hope you read on, but let us offer a few humble words. Of making many books there is no end, and a long preface is a chasing after wind. We pray you give us a moment's indulgence.

Our mission is to build a bridge into the past, before film, television, copyright, and internet swallowed up the world. Before 'content' was culture. If the reader finds friends from before the echo chamber, they may find armor and sword against the dreadful noise machine.

We are convinced that many authors and many books are ready to rise like Lazarus and reenter the world to remind the readers that their life has purpose; that their time should be valued; and their history is an honorable home.

This book was chosen because it is completely forgotten. It should be taught in middle school and high school as part of the American Canon. It is a glorious war story written by the son of one of our greatest Presidents. It includes a cameo from Lawrence of Arabia. For our nation to forget such a story shows how far we have fallen, how little we honor our military, our leaders, and our history and how little we expect of our leaders today.

We have reused the original photographs from the manuscript where possible, and replaced them where possible. We have included several other photographs from the Mesopotamia campaign, and other material such as newspaper clippings and soldier's poems.

Your Most Humble and Obedient Servant (YMHOS),
Arthur Bulkington,
Melville Bay

The War In

<u>Publisher's Foreword</u>

This story was one we had to work up to, to become worthy of resurrecting it and bringing it back into the world. We were surprised that we never heard about it before. We are disappointed it has not been made into a major motion picture. We may endeavor to create a script on this subject, simply because it is that good of a story. The image of an aeroplane airdropping a volume of Plutarch to Kermit Roosevelt in the Arabian Desert, is simply too perfect an image to remain forgotten by history. Kermit Roosevelt fought in Basra and Ramadi just like Jocko Willink did. When we found this, we felt as if we had swung the pick and struck gold or oil—and felt immensely enriched.

This is a combat memoir written by the son of Teddy Roosevelt, about his service in the middle east, in the First World War. He joined the British service because America had not joined the war. He had been married for only three years when he commissioned. His wedding in Spain in 1914 to the good lady Belle Willard received international newspaper coverage. He meets Lawrence of Arabia, an Australian poet named "Banjo," spies, Turks, Kurds, and sails with a squadron of the Japanese Navy. He races through the desert in custom Rolls Royce armored cars.

These are good reasons to read *The War in the Garden of Eden*. Reader, if you have found this, we assure you it is worth your time.

As we conceive it, the duty of a foreword is to make three points. For a text like this, it is almost unneccesary and remedial, but we owe it to the Reader who is on the fence.

1) Why we liked it. Why it was worth our time.

2) Why it has relevance today.

3) Why it is worth your precious reading time.

We liked the book because it felt like a discovery. It felt like finding a treasure chest in the sand. Or when the shovel dings against a forgotten king's tomb.

We liked it because it helped us remember what America was before the postwar amnesia copyright pocket universe world made us into slop consumers and doomscrollers.

We liked it simply because it is an adventure story, told well, that knows itself as part of the Western Canon, the American Story.

Kermit Roosevelt drops the names of books he read on nearly every other page. He quotes brief lines of poems as he narrates battle and travel. He saw himself as one man doing his duty, and living up to the standards of heroism and courage he read in Plutarch and Gibbon, in the Lusiad and Homer. We will remind the Reader that his brother Quentin Roosevelt died in the war, and this father, Colonel Roosevelt, would die shortly after the war. He suffered as much as any Argive ashore at Troy. And yet he has a sunny Greek courage that we don't see in our warfighters or politicans today.

A book like this justifies our entire venture. If we can help this book re-enter the larger culture, it could become a great miniseries or movie. It could be taught in schools. It could inspire a new generation of Americans to serve their country. It could inspire new leaders to live in a way that their sons choose to serve.

We suggest at least one important question brought up by this book is why is it not taught instead of, or alongside, the books we read about World War One in high school? Why don't we read this alongside *All Quiet on the Western Front* or *A Farewell to Arms*? Why isn't this book taught alongside books like *The Things We Carried*? Why don't we teach historical accounts of the war instead of novels?

The Reader likely understands that the answer to this question has much to do with the ideological currents that predominate in education and media production.

But even if the Reader understood this fact, they haven't been able to correct this one-sided worldview. Yes, Ernst Jünger wrote well about war, but we need an American voice.

We liked this book alot, and feel that it is the exact type of story we wish to resurrect and reinvite into the world. Kermit Roosevelt should be remembered, not just because of his father.

This is why it was worth our time to read, proofread, footnote, and add additional materials into the book you hold in your hand.

Why is this relevant today?

Our age has chosen to valorize the wrong types of people. The whole wave of 'representation' has given us a whole set of myths that have very little tie to the truth. You could cite movies and books and names by the dozen. Who invented the telescope, after all?[1]

Our age has chosen to demonize the Founders, Framers, and heritage stock of Americans. They have melted down statues and exhumed the graves of warriors to satiate their anti-American zeal.

So when we find a war story told by a President's son that noone has ever heard about, we have found the antidote to these evils. We have found a real 'hidden figure.'

The larger point is obvious. Many of our leaders and elected officials are guilty of being 'chickenhawks.' They have no 'skin in the game.' They are not tied by blood and honor to the fate of the nation, and make their choices based on careerist considerations. Those chickenhawks have come home to roost in 2025, as we are now on the brink of bankruptcy and World War Three.

The disconnect between leadership and military risk has caused the United States to make many obvious mistakes—in the Middle East sandbox. Turkey, Kurdistan, Iraq, Palestine and Persia remain geopolitically important to policymakers today.

It was once a privilege and an honor, as well as a duty, for the elites to serve their country on the battlefield. Theodore Roosevelt's sons all served in World War One. Kermit Roosevelt would serve in the Second World War as well. We deserve leaders who serve and whose children feel honor bound to serve and lead from the front.

---

1   We are referring to a now canonical tweet about the excesses of progressivism: "A black woman invented the telescope. You might disagree. You might even have evidence to the contrary. But you have to ask yourself: is this really worth losing my job over? A black woman invented the telescope."

This book is deeply relevant to our time. It is an antidote to bad history and anti-American ideology and historiography. It shows us what American leadership should see as its duty. It shows us how often we have made the same mistakes and fought the same wars.

We admit we have bordered on the hagiographic, and that our foreword isn't offering much pretense of neutrality, nor is it offering scholarly ballast to the narrative. We know that our Ideal Reader already has these values and some prior knowledge.

We believe that this means this book is worth the Reader's precious reading time.

The Reader wants books about heroes who know their duty to their Nation, their place in history, their role in upholding and adding to the Great Conversation that is the Western Canon.

The Reader wants to reconnect to real history—not advertising campaigns and screenwriter's fetishes.

The Reader wants men it can admire and honor and emulate.

If you agree with any of the above, then this book is worth your time. Join Kermit Roosevelt as he treks the desert, hoping for something good to read between battles.

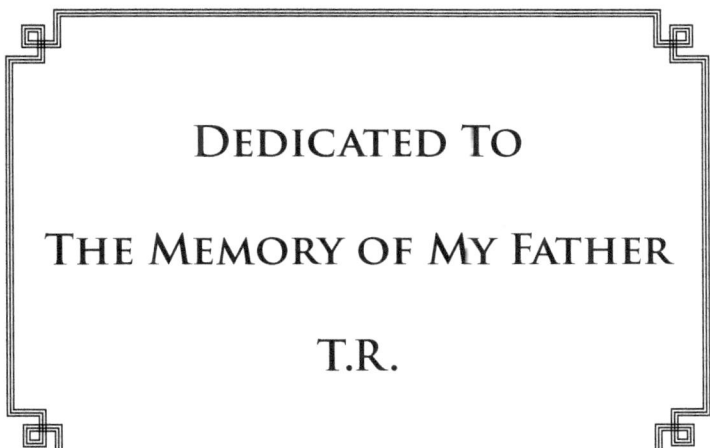

DEDICATED TO

THE MEMORY OF MY FATHER

T.R.

Newspaper Clippings

THE CALGARY DAILY HERALD

CALGARY, ALBERTA, SATURDAY, SEPTEMBER 27, 1919

Kermit Roosevelt is going to imitate his late lamented father by turning author. He has written an account of his war experiences under Generals Allenby and Maude in Palestin and Mesopotamia and has hit upon an attractive title for his book, "The War in the Garden of Eden."

MR. AND MRS. KERMIT ROOSEVELT AND THEIR WEDDING PARTY
Beside the Bride Is Her Sister, Miss Elizabeth Willard, and Behind Are
Col. Roosevelt, Ambassador Willard, Mrs. Willard, and Kermit
(Photo from Underwood & Underwood)

The War In

The Pueblo Indicator

June 6, 1914

KERMIT ROOSEVELT MEETS KING ALFONSO ANXIOUS TO MEET EX-PRESIDENT—

DATE OF WEDDING CHANGED

Madrid.—King Alfonso received in audience Kermit Roosevelt, together with Col. Joseph E. Willard, American ambassador to Spain; Mrs. Willard and Miss Belle Wyatt Willard.

His majesty conversed for some time with Kermit Roosevelt, questioning him on his recent experiences in Brazil and listening attentively to the narrative of the explorations made by the party.

The king said he desired to meet Col. Roosevelt when he came to Spain to attend the wedding.

The church wedding as well as the civil wedding is to take place on June 10 instead of June 11, owing to the latter being Corpus Christi day.

New York.—Theodore Roosevelt, accompanied by Philip Roosevelt, a young cousin, and his eldest daughter, Mrs. Nicholas Longworth of Cincinnati, sailed for Spain on the steamship Olympia to attend the wedding of his son, Kermit, in Madrid on June 10, to Miss Belle Willard.

Pittsburgh Press

June 11, 1914

BRILLIANT GATHERING AT THE ROOSEVELT-WILLARD WEDDING

Madrid. June 11—A brilliant gathering comprising the elite of Spanish officialdom and aristocracy, together with practically all the members of the diplomatic corps and their ladies, witnessed the marriage of Miss Belle Wyatt Willard, daughter of Joseph P. Willard, American ambassador to Spain, and Kermit Roosevelt, son of Col. Theodore Roosevelt, ex-president of the United States. The bridgegroom's father came over from New York to attend the wedding. The ceremony, which took place at noon in the British Embassy chapel, was performed by the Rev. Dr. Watson, rector of the American church in Paris, assisted by the Rev. Herbert Brown, chaplain of the British Embassy in Madrid. The bride was attended by her sister, Miss Elizabeth Willard, as maid of honor, and by the following bridesmaids: Her Serene Highness the Princess Pella of Thurn of Taxis, daughter of Prince Ratibor, the German Ambassador in Madrid; Miss Katherine Page, daughter of the American Ambassador in London; Mlle. Gilone le Venour de Tillieres of Paris and Miss Virginia Christian of Richmond, VA. Following a brief honeymoon on the continent, the young couple will leave for Brazil to establish their future home at San Paulo, where Kermit Roosevelt is engaged in the railroad business.

The bride, now 21, was one of the most attractive girls in the younger social set at Richmond, the family home of the Willards, where she made her debut two years ago. While en route with her mother and younger sister last February to join her father at his new post she was presented at the Court of St. James in London, and upon her arrival here she quickly became a general favorite in Spanish society. Kermit Roosevelt, 24 years old, has seen considerably more of the world than usually falls to the lot of young men of his age. After returning from the big game hunt in Africa with his father in 1910 he went out west on a mountain sheep shooting expedition. Following his graduation from Harvard in 1912 he made a visit in England and later took up railroading in South America. Last winter he accompanied Col. Roosevelt on his hunting, exploring and river-discovering trip in the Amazon country, coming direct from Brazil to this city, where he arrived three weeks ago. Kermit met his future wife through the good offices of his sister. Mrs. Richard Derby then Miss Ethel Roosevelt, who had become a great chum of Miss Willard's during a sojourn at Hot Springs, VA. Ethel took her new friend to Oyster Bay, where Kermit speedily fell a victim to her charms.

The Evening Record
Windsor, Ontario
August 11, 1917
ROOSEVELT'S SON, CANADIAN CAPTAIN, VISITS "SAMMIES" YOUNG OFFICER IS EN ROUTE TO MESOPOTAMIA WAR FRONT

By J.W. Pegler. American Field Headquarters, August 4.

Kermit Roosevelt, now a captain in the Canadian army, was a visitor at the training camp of the American contingent today. Young Roosevelt is enroute to Mesopotamia. He saw the Sammies go through their various war practice stunts.

General Sibert plans to hold the first review of the American contingent within a few days. The Sammies will march by their commanders over a parade ground laid out on a great plain.

The American camp commander is greatly pleased with the progress his men have made. He will depart soon for a tour of inspection of the French front.

Through yesterday a sharp lookout was kept by all for aeroplanes. This followed the sounding of an alarm from headquarters on Sunday, when everyone in camp and in the town nearby rushed into the streets in search of the hostile aircraft believed to be approaching.

Supply difficulties are now disappearing and life in the camp is rapidly becoming more comfortable for the Sammies. A candy and tobacco store, mounted on an automobile, is soon to begin a daily tour of the camp.

The Pittsburgh Gazette Times

April 28, 1918

KERMIT ROOSEVELT APPOINTED FIELD ARTILLERY CAPTAIN

ALL OF COLONEL'S SONS NOW IN THE UNITED STATES FIGHTING FORCES

Washington, April 27.—Announcement was made today by the War Department today that Kermit Roosevelt, second son of Col. Theodore Roosevelt, had been appointed a captain of field artillery in the National Army. Capt. Roosevelt has been, or is still, serving with the British Expeditionary Forces in Mesopotamia as a captain of sappers in the British Army and has seen much active service. His appointment to a captaincy in the National Army brings all of Col. Roosevelt's sons into the military service of the United States.

Theodore Roosevelt, Jr., the eldest son of the former President, is a major of infantry now on the fighting line in France. The third son, Archibald Roosevelt, was wounded in France recently, and was promoted from first lieutenant to captain of infantry. He is in a hospital recovering from his wounds.

Quentin Roosevelt, the youngest son, who was a baby when Col. Roosevelt became President, is a member of the American aviation corps in France. He has been ill recently with pneumonia. Col. Roosevelt's son-in-law, Dr. Richard Dreby, who married Miss Ethel Roosevelt, is a major of the United States Medical Reserve Corps, and is in France, where his wife is also doing war work.

The Washington Reporter

WASHINGTON, PA., WEDNESDAY, JULY 11, 1917.

KERMIT ROOSEVELT JOINS BRITISH IN MESOPOTAMIA

Plattsburg, N. Y., July 11.—Kermit Roosevelt has received a cablegram containing an offer, which he has accepted, of a staff commission with the British army operating against the Turks in Asia Minor. He was granted his discharge from the officers' training camp here, and, accompanied by his wife, left for Oyster Bay to join his father before sailing on Saturday for Spain.

The Evening Independent

Editors and Owners. ISSUED EVERY AFTERNOON (EXCEPT SUNDAY) AT 4:30 O'CLOCK FROM THE INDEPENDENT BUILDING. (Delivered Two Weeks

ST. PETERSBURG, PINELLAS COUNTY, FLORIDA, TUESDAY, AUGUST 27, 1918

MILITARY SERVICE CROSS AWARDED KERMIT ROOSEVELT

London, Aug. 27.—The award has been made to temporary and honorary Captain Kermit Roosevelt of the military cross for service in Mesopotamia last night. Until he joined the American forces he was attached to the British army in that region on special duty.

Aviator Quentin Roosevelt Killed In Air Battle

NOON EXTRA

Buy Thrift Stamps Now

War Savings Stamps Issued By U. S.

Issued By The
Portsmouth
Daily Times

ELEVEN O'CLOCK PORTSMOUTH, OHIO WEDNESDAY, JULY 17, 1918. (Established April 30, 1853) PRICE ONE CENT

Quentin Roosevelt Is Killed

PARIS, July 17.—Lieut. Quentin Roosevelt, youngest son of the former President of the United States has been killed in an air fight, the semi-official Havas News Agency announces. His machine fell into the enemy lines.

LONDON, July 17.—Lieut. Quentin Roosevelt, Col. Theodore Roosevelt's youngest son, who has been attached to the American line forces on the Marne front, was killed at Chateau Thierry on July 14, says a dispatch from Paris to the Exchange Telegraph Company.

Lieut. Roosevelt, the dispatch says, was returning from a patrol fight when he was attacked by a German squadron.

It was seen that Roosevelt suddenly lost control of his machine, having probably received a mortal wound.

PARIS, July 17.—Lieut. Quentin Roosevelt, of the American air service, youngest son of former President Theodore Roosevelt, is missing. His machine was seen to fall within the German lines. It was not in flames when it fell.

Philip Roosevelt, Quentin's cousin, witnessed the air battle in the vicinity of Chateau Thierry, in which Quentin was engaged and saw the machine fall but did not know until later that the airplane was that of his cousin, Le Journal says today.

U. S. Troops Advance Against Strong Machine Gun Fire Over Ground Covered With Hun Dead

The Portsmouth Daily Times.

VOLUME TWENTY-FIVE   14 PAGES TODAY   PORTSMOUTH, OHIO,   SATURDAY, JULY 20, 1918.   Price, Single Copy Three Cents

THEODORE ROOSEVELT, JR.
IS SLIGHTLY WOUNDED

OYSTER BAY, N. Y., July 20.—Major Theodore Roosevelt, Jr., has been slightly wounded and taken to a hospital in Paris, according to a cable message received tonight by his father, Colonel Theodore Roosevelt, from his daughter-in-law, Mrs. Theodore Roosevelt, Jr.

The cablegram was sent from Paris where Mrs. Roosevelt is in a Red Cross hospital. It read:

"Ted wounded. Not seriously. Here with me. Not any danger. No cause for anxiety."

Major Roosevelt was recently cited for gallantry after having been gassed about three weeks ago.

The news of Major Roosevelt having been wounded followed immediately the report from Paris that German aviators had dropped a note behind the Allied lines confirming fears of the death of Lieutenant Quentin Roosevelt in an aerial engagement. Captain Archie Roosevelt, another of the former President's sons, is now recovering from severe wounds. Kermit Roosevelt, a fourth son who served as captain in the British army in Mesopotamia was recently appointed a captain in the United States army and is now reported on his way to France. He was decorated by the British with the military cross for gallantry in action.

The War In

The Garden of Eden

<u>Campaign Maps</u>

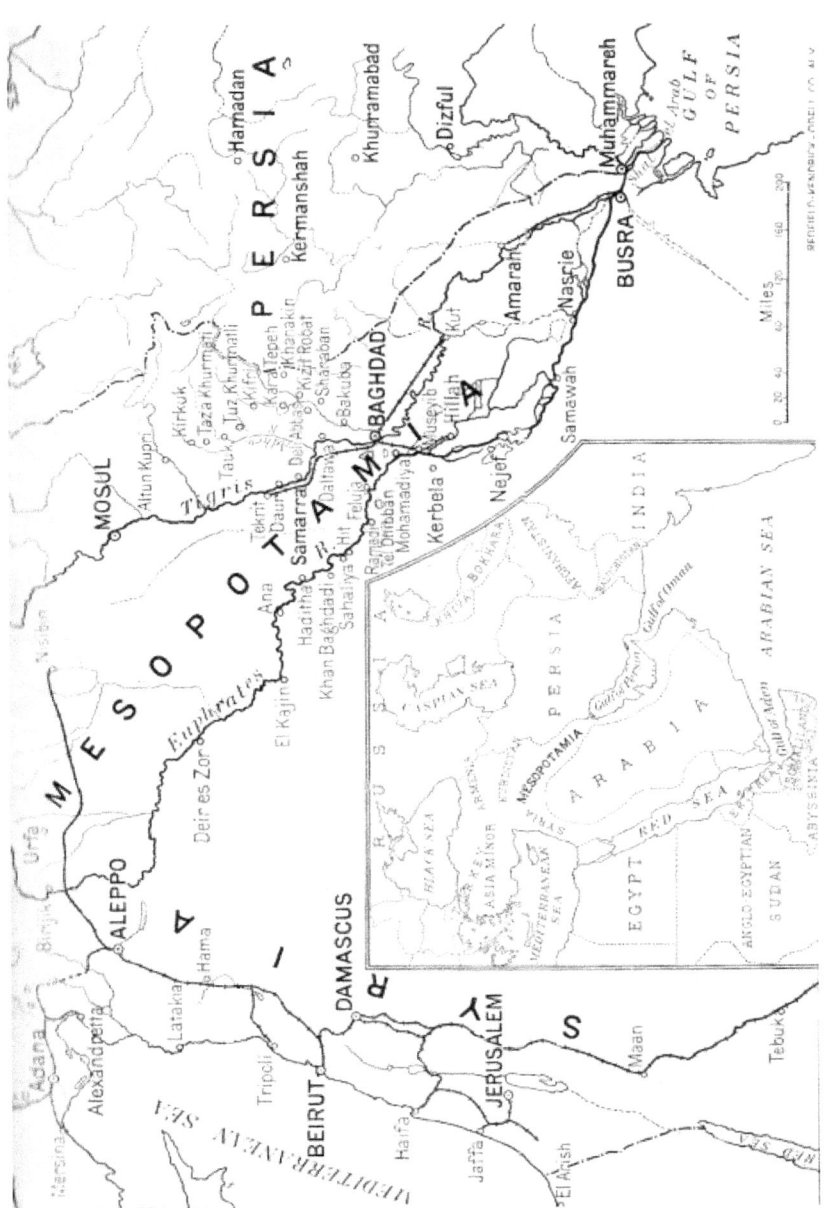

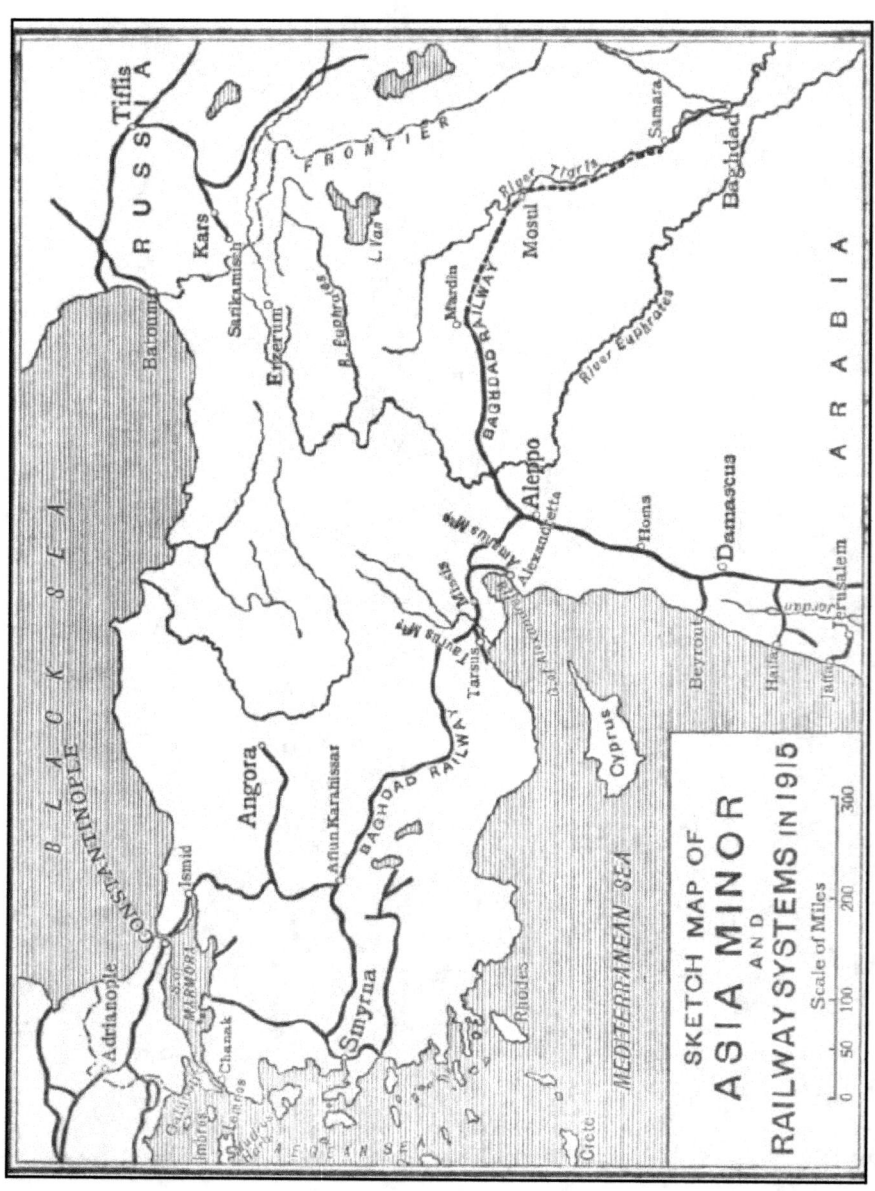

SKETCH MAP OF
ASIA MINOR
AND
RAILWAY SYSTEMS IN 1915

Scale of Miles

0   50   100   200   300

Map of Baghdad and Ramadi

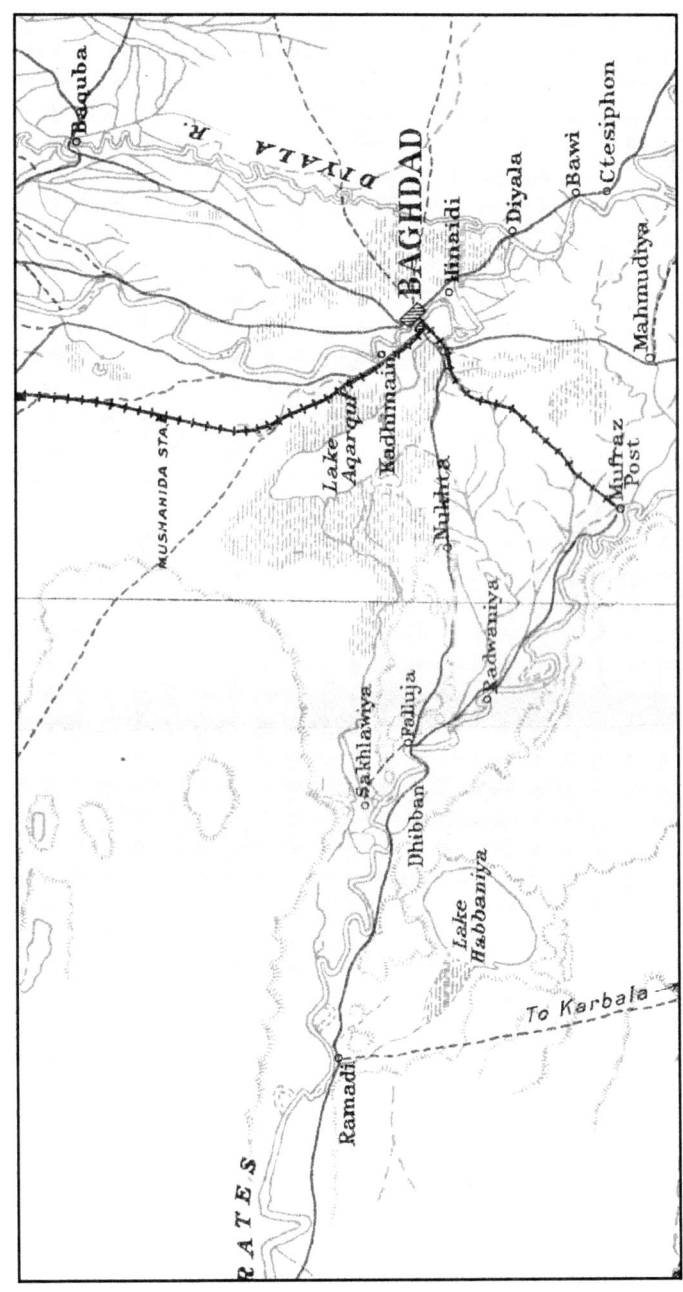

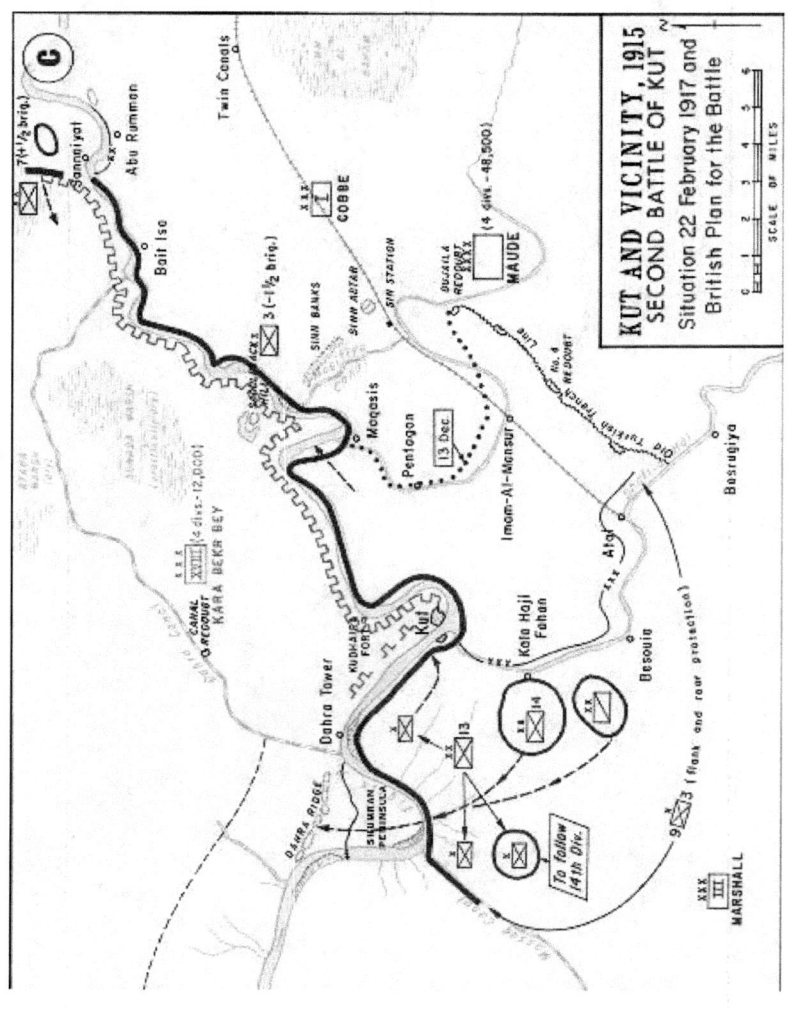

Map of Ramadi Area

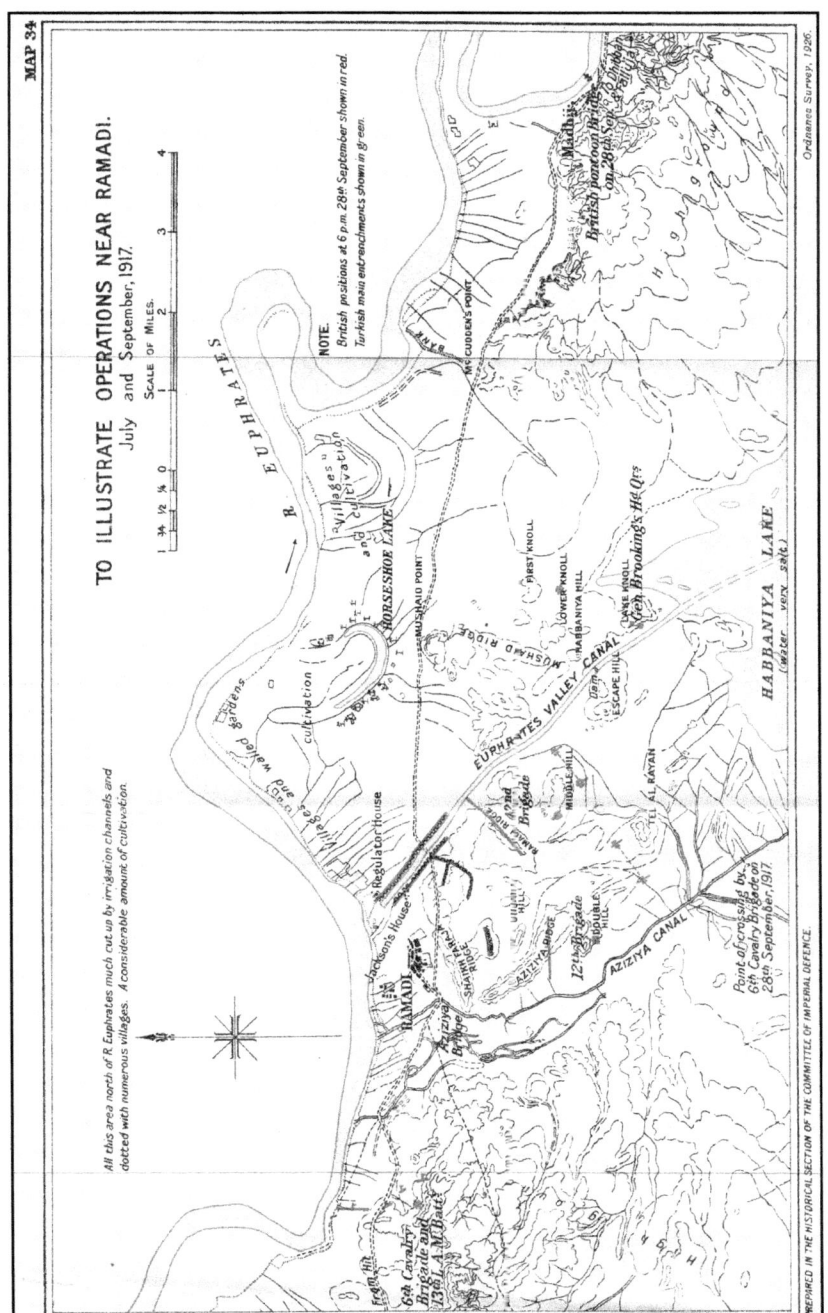

Map of Tikrit and Daur

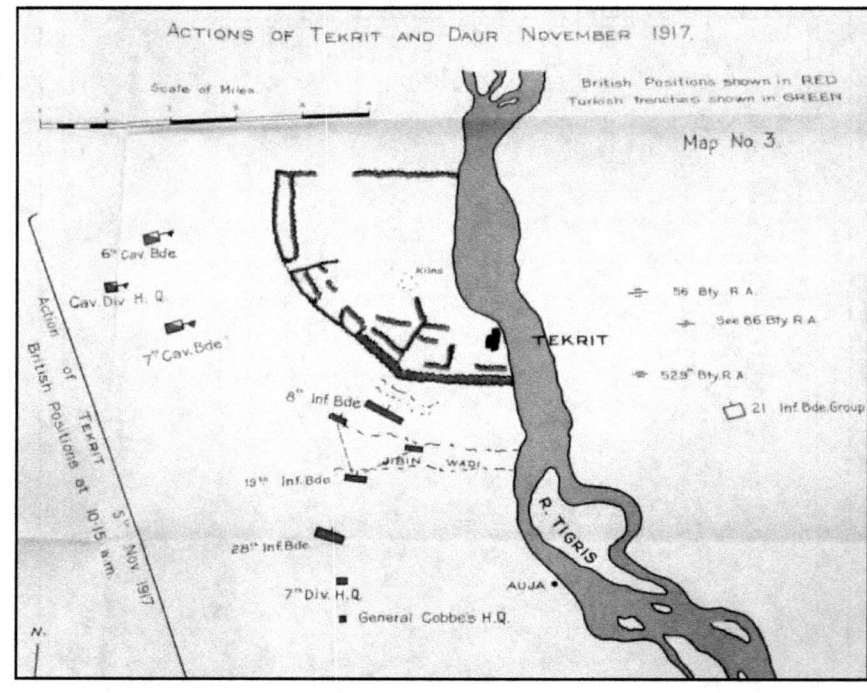

To My Camel
by Major Oliver Hogue, Imperial Camel Corps

You're an ugly smellful creature;
You're a blot upon the plain;
I have seen Mohamed beat you,
And it gave me little pain.
You're spiteful and you're lazy,
You'd send a white man crazy,
But I reckon you're a daisy
When the Turks come out again.

Your head is most unsightly,
And so is your humpy back;
I hear you roaring nightly,
When you're loading for the track.
You're bow-legged and you're bandy,
But in this desert sandy
It's as well to have you handy:
You're a mighty useful hack.

You shake me something cruel
When you try to do a trot;
I've got to take my gruel,
But you make it very hot:
I've somehow got a notion
That your humpty-dumpty motion
Is worse than on the ocean,
It's a nasty way you've got.

It's a sun-scorched land, the East is.
So we need you when we trek.
My old prad a better beast is,
But he'd soon become a wreck.
You thirst a week unblinking,
And when I see you drinking,
You always get me thinking:
Lord, I wish I had your neck.

In Defence of the Bush
by A.B. Paterson

So you're back from up the country, Mister Lawson, where you went,
    And you're cursing all the business in a bitter discontent;
Well, we grieve to disappoint you, and it makes us sad to hear
    That it wasn't cool and shady — and there wasn't whips of beer,
And the looney bullock snorted when you first came into view —
    Well, you know it's not so often that he sees a swell like you;
And the roads were hot and dusty, and the plains were burnt and brown,
    And no doubt you're better suited drinking lemon-squash in town.
Yet, perchance, if you should journey down the very track you went
    In a month or two at furthest, you would wonder what it meant;
Where the sunbaked earth was gasping like a creature in its pain
    You would find the grasses waving like a field of summer grain,
And the miles of thirsty gutters, blocked with sand and choked with mud,
    You would find them mighty rivers with a turbid, sweeping flood.
For the rain and drought and sunshine make no changes in the street,
    In the sullen line of buildings and the ceaseless tramp of feet;
But the bush has moods and changes, as the seasons rise and fall,
And the men who know the bush-land — they are loyal through it all.

. . . . .

But you found the bush was dismal and a land of no delight —
    Did you chance to hear a chorus in the shearers' huts at night?
Did they 'rise up William Riley' by the camp-fire's cheery blaze?
    Did they rise him as we rose him in the good old droving days?
And the women of the homesteads and the men you chanced to meet —
    Were their faces sour and saddened like the 'faces in the street'?
And the 'shy selector children' — were they better now or worse
    Than the little city urchins who would greet you with a curse?
Is not such a life much better than the squalid street and square
    Where the fallen women flaunt it in the fierce electric glare,
Wher the sempstress plies her needle till her eyes are sore and red
    In a filthy, dirty attic toiling on for daily bread?
Did you hear no sweeter voices in the music of the bush
    Than the roar of trams and buses, and the war-whoop of 'the push'?
Did the magpies rouse your slumbers with their carol sweet and strange?
    Did you hear the silver chiming of the bell-birds on the range?
But, perchance, the wild birds' music by your senses was despised,
    For you say you'll stay in townships till the bush is civilized.
Would you make it a tea-garden, and on Sundays have a band
Where the 'blokes' might take their 'donahs', with a 'public' close at hand?
You had better stick to Sydney and make merry with the 'push',
    For the bush will never suit you, and you'll never suit the bush

CHAPTER ONE
OFF FOR MESOPOTAMIA

It was at Taranto that we embarked for Mesopotamia. Reinforcements were sent out from England in one of two ways—either all the way round the Cape of Good Hope, or by train through France and Italy down to the desolate little seaport of Taranto, and thence by transport over to Egypt, through the Suez Canal, and on down the Red Sea to the Indian Ocean and the Persian Gulf. The latter method was by far the shorter, but the submarine situation in the Mediterranean was such that convoying troops was a matter of great difficulty. Taranto is an ancient Greek town, situated at the mouth of a landlocked harbor, the entrance to which is a narrow channel, certainly not more than two hundred yards across.'The old part of the town is built on a hill, and the alleys and runways winding among the great stone dwellings serve as streets. As is the case with maritime towns, it is along the wharfs that the most interest centres. During one afternoon I wandered through the old town and listened to the fisherfolk singing as they overhauled and mended their nets. Grouped around a stone archway sat six or seven women and girls. They were evidently members of one family—a grandmother, her daughters, and their children. The old woman, wild, dark, and hawk-featured, was blind, and as she knitted she chanted some verses. I could only understand occasional words and phrases, but it was evidently a long epic. At intervals her listeners would break out in comments as they worked, but, like "Othere, the old sea-captain," she "neither paused nor stirred."[1]

There are few things more desolate than even the best situated "rest-camps"—the long lines of tents set out with military precision, the trampled grass, and the board walks; but the one at Taranto where we awaited embarkation was peculiarly dismal even

1   Henry Wadsworth Longfellow, *The Discoverer of the North Cape - A Leaf from King Alfred's Orosius,* "But Othere, the old sea-captain,/He neither paused nor stirred,/Till the King listened, and then/Once more took up his pen,/And wrote down every word.

for a rest-camp. So it happened that when Admiral Mark Kerr,[2] the commander of the Mediterranean fleet, invited me to be his guest aboard H.M.S. *Queen* until the transport should sail, it was in every way an opportunity to be appreciated.

In the British Empire the navy is the "senior service," and I soon found that the tradition for the hospitality and cultivation of its officers was more than justified. The admiral had travelled, and read, and written, and no more pleasant evenings could be imagined than those spent in listening to his stories of the famous writers, statesmen, and artists who were numbered among his friends. He had always been a great enthusiast for the development of aerial warfare, and he was recently in Nova Scotia in command of the giant Handley-Page machine which was awaiting favorable weather conditions in order to attempt the nonstop transatlantic flight. Among his poems' stands out the "Prayer of Empire," which, oddly enough, the former German Emperor greatly admired, ordering it distributed throughout the imperial navy! The Kaiser's feelings toward the admiral have suffered an abrupt change, but they would have been even more hostile had England profited by his warnings:

> *"There's no menace in preparedness, no threat in being strong,*
> *If the people's brain be healthy and they think no thought of wrong."[3]*

After four or five most agreeable days aboard the *Queen* the word came to embark, and I was duly transferred to the *Saxon*, an old Union Castle liner that was to run us straight through to Busra.

As we steamed out of the harbor we were joined by two diminutive Japanese destroyers which were to convoy us. The menace of the submarine being particularly felt in the Adriatic, the transports travelled only by night during the first part of the voyage. To a landsman it was incomprehensible how it was possible for us to pursue our zigzag course in the inky blackness and avoid collisions, particularly when it was borne in mind that our ship was English

---

2   Admiral Mark Kerr, CB, CVO, (1864-1944) Served in the Royal Navy then the RAF.

3   Poem by the Admiral. The full text of the poem 'Prayer of Empire' could not be located, but he wrote at least one book of verse, *The Destroyer and a Cargo of Notions*, two memoirs, and two studies on naval history.

and our convoyers were Japanese. During the afternoon we were drilled in the method of abandoning ship, and I was put in charge of a lifeboat and a certain section of the ropes that were to be used in our descent over the side into the water. Between twelve and one o'clock that night we were awakened by three blasts, the preconcerted danger-signal. Slipping into my life-jacket, I groped my way to my station on deck. The men were filing up in perfect order and with no show of excitement. A ship's officer passed and said he had heard that we had been torpedoed and were taking in water. For fifteen or twenty minutes we knew nothing further. A Scotch captain who had charge of the next boat to me came over and whispered: "It looks as if we'd go down. I have just seen a rat run out along the ropes into my boat!" That particular rat had not been properly brought up, for shortly afterward we were told that we were not sinking. We had been rammed amidships by one of the escorting destroyers, but the breach was above the water-line. We heard later that the destroyer, though badly smashed up, managed to make land in safety.

We laid up two days in a harbor on the Albanian coast, spending the time pleasantly enough in swimming and sailing, while we waited for a new escort. Another night's run put us in Navarino Bay. The grandfather of Lieutenant Finch Hatton, one of the officers on board, commanded the Allied forces in the famous battle fought here in 1827, when the Turkish fleet was vanquished and the independence of Greece assured.[4]

Several days more brought us to Port Said, and after a short delay we pushed on through the canal and into the Red Sea. It was August, and when one talks of the Red Sea in August there is no further need for comment. The *Saxon* had not been built for the tropics. She had no fans, nor ventilating system such as we have on the United Fruit boats. Some unusually intelligent stokers had deserted at Port Said, and as we were in consequence short-handed, it was suggested that any volunteers would be given a try. Finch Hatton and I felt that our years in the tropics should qualify us,

---

4   Denys Finch-Hatton, MC (1887-1931) was a big game hunter, photographer, pilot, and womanizer. Inspiration for the main character in the novel *Out of Africa* by Isak Dinesen. Played by Robert Redford in the 1985 film.

and that the exercise would improve our dispositions. We got the exercise. Never have I felt anything as hot, and I have spent August in Yuma, Arizona, and been in Italian Somaliland and the Amazon Valley. The shovels and the handles of the wheelbarrows blistered our hands.

We had a number of cases of heat-stroke, and the hospital facilities on a crowded transport can never be all that might be desired. The first military burial at sea was deeply impressive. There was a lane of Tommies drawn up with their rifles reversed and heads bowed; the short, classic burial service was read, and the body, wrapped in the Union Jack, slid down over the stern of the ship. Then the bugles rang out in the haunting, mournful strains of the "Last Post," and the service ended with all singing "Abide With Me."[5]

We sweltered along down the Red Sea and around into the Indian Ocean. We wished to call at Aden in order to disembark some of our sick, but were ordered to continue on without touching. Our duties were light, and we spent the time playing cards and reading. The Tommies played "house" from dawn till dark. It is a game of the lotto variety. Each man has a paper with numbers written on squares; one of them draws from a bag slips of paper also marked with numbers, calls them out, and those having the number he calls cover it, until all the numbers on their paper have been covered. The first one to finish wins, and collects a penny from each of the losers. The caller drones out the numbers with a monotony only equalled by the brain-fever bird, and quite as disastrous to the nerves. There are certain conventional nicknames: number one is always "Kelley's eye," eleven is "legs eleven," sixty-six is "clickety click," and the highest number is "top o' the 'ouse." There is another game that would be much in vogue were it not for the vigilance of the officers. It is known as "crown and anchor," and the advantage lies so strongly in favor of the banker that he cannot fail to make a good income, and therefore the game is forbidden under the severest penalties.

5　　"Last Post" is a British and Commonwealth bugle call for military funerals and commemorations of war dead. "Abide with Me" is a Christian hymn by Henry Francis Lye written in 1847.

As we passed through the Strait of Ormuz memories of the early days of European supremacy in the East crowded back, for I had read many a vellum-covered volume in Portuguese about the early struggles for supremacy in the gulf. One in particular interested me. The Portuguese were hemmed in at Ormuz by a greatly superior English force. The expected reinforcements never arrived, and at length their resources sank so low, and they suffered in addition, or in consequence, so greatly from disease that they decided to sail forth and give battle. This they did, but before they joined in fight the ships of the two admirals sailed up near each other—the Portuguese commander sent the British a gorgeous scarlet ceremonial cloak, the British responded by sending him a handsomely embossed sword. The British admiral donned the cloak, the Portuguese grasped the sword; a page brought each a cup of wine; they pledged each other, threw the goblets into the sea, and fell to. The British were victorious. Times indeed have sadly changed in the last three hundred years!

I was much struck with the accuracy of the geographical descriptions in Camoens' letters and odes. He is the greatest of the Portuguese poets and wrote the larger part of his master-epic, "*The Lusiad*," while exiled in India.[6] For seventeen years he led an adventurous life in the East; and it is easy to recognize many harbors and stretches of coast line from his inimitable portrayal.

Busra, our destination, lies about sixty miles from the mouth of the Shatt el Arab, which is the name given to the combined Tigris and Euphrates after their junction at Kurna, another fifty or sixty miles above. At the entrance to the river lies a sand-bar, effectively blocking access to boats of as great draft as the *Saxon*. We therefore transshipped to some British India vessels, and exceedingly comfortable we found them, designed as they were for tropic runs. We steamed up past the Island of Abadan, where stand the refineries of the Anglo-Persian Oil Company. It is hard to overestimate the important part that company has played in the conduct of the Mesopotamian campaign. Motor transport was nowhere else a greater

---

6   Kermit Roosevelt alludes to this book several times. It is an epic poem about the voyages of Vasco de Gama, written by Luís Vaz de Camões(1525-1580), considered the national epic of Portugal.

necessity. There was no possibility of living on the country; at first, at all events. General Dickson, the director of local resources, later set in to so build up and encourage agriculture that the army should eventually be supported, in the staples of life, by local produce. Transportation was ever a hard nut to crack. Railroads were built, but though the nature of the country called for little grading, obtaining rails, except in small quantities, was impossible. The ones brought were chiefly secured by taking up the double track of Indian railways. This process naturally had a limit, and only lines of prime importance could be laid down. Thus you could go by rail from Busra to Amara, and from Kut to Baghdad, but the stretch between Amara and Kut had never been built, up to the time I left the country. General Maude once told me that pressure was being continually brought by the high command in England or India to have that connecting-link built, but that he was convinced that the rails would be far more essential elsewhere, and had no intention of yielding.

I don't know the total number of motor vehicles, but there were more than five thousand Fords alone. On several occasions small columns of infantry were transported in Fords, five men and the driver to a car. Indians of every caste and religion were turned into drivers, and although it seemed sufficiently out of place to come across wizened, khaki-clad Indo-Chinese driving lorries in France, the incongruity was even more marked when one beheld a great bearded Sikh with his turbaned head bent over the steering-wheel of a Ford.

Modern Busra stands on the banks of Ashar Creek. The ancient city whence Sinbad the sailor set forth is now seven or eight miles inland, buried under the shifting sands of the desert. Busra was a seaport not so many hundreds of years ago. Before that again, Kurna was a seaport, and the two rivers probably only joined in the ocean, but they have gradually enlarged the continent and forced back the sea. The present rate of encroachment amounts, I was told, to nearly twelve feet a year.

The modern town has increased many fold with the advent of the Expeditionary Force, and much of the improvement is of a necessarily permanent nature; in particular the wharfs and roads. Indeed, one of the most striking features of the Mesopotamian campaign is the permanency of the improvements made by the British. In order to conquer the country it was necessary to develop it,—build railways and bridges and roads and telegraph systems,—and it has all been done in a substantial manner. It is impossible to contemplate with equanimity the possibility of the country reverting to a rule where all this progress would soon disappear and the former stagnancy and injustice again hold sway.

Ashar Creek at Busra

Indian Troops landing in Basra

As soon as we landed I wandered off to the bazaar—"suq" is what the Arab calls it. In Busra there are a number of excellent ones. By that I don't mean that there are art treasures of the East to be found in them, for almost everything could be duplicated at a better price in New York. It is the grouping of wares, the mode of sale, and, above all, the salesmen and buyers that make a bazaar— the old bearded Persian sitting cross-legged in his booth, the motley crowd jostling through the narrow, vaulted passageway, the veiled women, the hawk-featured, turbaned men, the Jews, the Chaldeans, the Arabs, the Armenians, the stalwart Kurds, and through it all a leaven of khaki-clad Indians, purchasing for the regimental mess. All these and an ever-present exotic, intangible something are what the bazaar means. Close by the entrance stood a booth festooned with lamps and lanterns of every sort, with above it scrawled "Aladdin-Ibn-Said." My Arabic was not at that time sufficient to enable me to discover from the owner whether he claimed illustrious ancestry or had merely been named after a patron saint.

A few days after landing at Busra we embarked on a paddle-wheel boat to pursue our way up-stream the five hundred intervening miles to Baghdad. Along the banks of the river stretched endless miles of date-palms. We watched the Arabs at their work of fertilizing them, for in this country these palms have to depend on human agency to transfer the pollen. At Kurna we entered the Garden of Eden, and one could quite appreciate the feelings of the disgusted Tommy who exclaimed: "If this is the Garden, it wouldn't take no bloody angel with a flaming sword to turn me back."[7] The direct descendant of the Tree is pointed out; whether its properties are inherited I never heard, but certainly the native would have little to learn by eating the fruit.

Above Kurna the river is no longer lined with continuous palm-groves; desert and swamps take their place—the abode of the amphibious, nomadic, marsh Arab. An unruly customer he is apt to prove himself, and when he is "wanted" by the officials, he retires to his watery fastnesses, where he can remain in complete safety unless betrayed by his comrades. On the banks of the Tigris stands

---

7   Genesis 3:24. "So he drove out the man; and he placed at the east of the garden of Eden Cherubims, and a flaming sword which turned every way, to keep the way of the tree of life."

Ezra's tomb. It is kept in good repair through every vicissitude of rule, for it is a holy place to Moslem and Jew and Christian alike.

The third night brought us to Amara. The evening was cool and pleasant after the scorching heat of the day, and Finch Hatton and I thought that we would go ashore for a stroll through the town. As we proceeded down the bank toward the bridge, I caught sight of a sentry walking his post. His appearance was so very important and efficient that I slipped behind my companion to give him a chance to explain us. "Halt! Who goes there?" "Friend," replied Finch Hatton. "Advance, friend, and give the countersign." F. H. started to advance, followed by a still suspicious me, and rightly so, for the Tommy, evidently member of a recent draft, came forward to meet us with lowered bayonet, remarking in a business-like manner: "There isn't any countersign."

Except for the gunboats and monitors all river traffic is controlled by the Inland Water Transport Service. The officers are recruited from all the world over. I firmly believe that no river of any importance could be mentioned but what an officer of the I. W. T. could be found who had navigated it. The great requisite for transports on the Tigris was a very light draft, and to fill the requirements boats were requisitioned ranging from penny steamers of the Thames to river-craft of the Irrawaddy. Now in bringing a penny steamer from London to Busra the submarine is one of the lesser perils, and in supplying the wants of the Expeditionary Force more than eighty vessels were lost at sea, frequently with all aboard.

As was the custom, we had a barge lashed to either side. These barges are laden with troops, or horses, or supplies. In our case we had the first Bengal regiment—a new experiment, undertaken for political reasons. The Bengali is the Indian who most readily takes to European learning. Rabindranath Tagore is probably the most widely known member of the race. They go to Calcutta University and learn a smattering of English and absorb a certain amount of undigested general knowledge and theory. These partially educated Bengalis form the Babu class, and many are employed in the railways. They delight in complicated phraseology, and this coupled with their accent and seesaw manner of speaking supply the English a constant source of caricature. As a race they are inclined to be vain and boastful, and are ever ready to nurse a grievance against the British Government, feeling that they have been provided with an education but no means of support. The government felt that it might help to calm them if a regiment were recruited and sent to Mesopotamia. How they would do in actual fighting had never been demonstrated up to the time I left the country, but they take readily to drill, and it was amusing to hear them ordering each other about in their clipped English. They were used for garrisoning Baghdad.

After we left Amara we continued our winding course upstream. A boat several hours ahead may be seen only a few hundred yards distant across the desert. The banks are so flat and level that it looks as if the other vessels were steaming along on land. The Arab river-craft was most picturesque. At sunset a mahela, bearing down with filled sail, might have been the model for Maxfield Parrish's *Pirate Ship*.[8] The Arab women ran along the bank beside us, carrying baskets of eggs and chickens, and occasionally melons. They were possessed of surprising endurance, and would accompany us indefinitely, heavily laden as they were. Their robes trailed in the wind as they jumped ditches, screaming out their wares without a moment's pause. An Indian of the boat's crew was haggling with a woman about a chicken. He threw her an eight-anna piece. She picked up the money but would not hand him the chicken, holding out for her original price. He jumped ashore, intending to take the chicken. She had a few yards' start and made the most of it. In and out they chased, over hedge and ditch, down the bank and up again. Several times he almost had her. She never for a moment ceased screeching—an operation which seemed to affect her wind not a particle. At the end of fifteen minutes the Indian gave up amid the delighted jeers of his comrades, and returned shamefaced and breathless to jump aboard the boat as we bumped against the bank on rounding a curve.

One evening we halted where, not many months before, the last of the battles of Sunnaiyat had been fought. There for months the British had been held back, while their beleaguered comrades

8 Maxfield Parrish (1870-1966) painter. The painting is exactly as he describes.

10

in Kut could hear the roar of the artillery and hope against hope for the relief that never reached them. It was one phase of the campaign that closely approximated the gruelling trench warfare in France. The last unsuccessful attack was launched a week before the capitulation of the garrison, and it was almost a year later before the position was eventually taken. The front-line trenches were but a short distance apart, and each side had developed a strong and elaborate system of defense. One flank was protected by an impassable marsh and the other by the river. When we passed, the field presented an unusually gruesome appearance even for a battle-field, for the wandering desert Arabs had been at work, and they do not clean up as thoroughly as the African hyena. A number had paid the penalty through tampering with unexploded grenades and "dud" shells, and left their own bones to be scattered around among the dead they had been looting. The trenches were a veritable Golgotha with skulls everywhere and dismembered legs still clad with puttees and boots.

At Kut we disembarked to do the remaining hundred miles to Baghdad by rail instead of winding along for double the distance by river, with a good chance of being hung up for hours, or even days, on some shifting sand-bar. At first sight Kut is as unpromising a spot as can well be imagined, with its scorching heat and its sand and the desolate mud-houses, but in spite of appearances it is an important and thriving little town, and daily becoming of more consequence.

The railroad runs across the desert, following approximately the old caravan route to Baghdad. A little over half-way the line passes the remaining arch of the great hall of Ctesiphon. This hall is one hundred and forty-eight feet long by seventy-six broad. The arch stands eighty-five feet high. Around it, beneath the mounds of desert sand, lies all that remains of the ancient city. As a matter of fact the city is by no means ancient as such things go in Mesopotamia, dating as it does from the third century B. C., when it was founded by the successors of Alexander the Great.[9]

My first night in Baghdad I spent in General Maude's house,

9   Alexander the Great founded the city, and named it after the orator Ctesiphon, known for his dispute with Demosthenes.

on the river-bank. The general was a striking soldierly figure of a man, standing well over six feet. His military career was long and brilliant. His first service was in the Coldstream Guards. He distinguished himself in South Africa. Early in the present war he was severely wounded in France. Upon recovering he took over the Thirteenth Division, which he commanded in the disastrous Gallipoli campaign, and later brought out to Mesopotamia. When he reached the East the situation was by no means a happy one for the British. General Townshend was surrounded in Kut, and the morale of the Turk was excellent after the successes he had met with in Gallipoli. In the end of August, 1916, four months after the fall of Kut, General Maude took over the command of the Mesopotamian forces. On the 11th of March of the following year he occupied Baghdad, thereby re-establishing completely the British prestige in the Orient. One of Germany's most serious miscalculations was with regard to the Indian situation. She felt confident that, working through Persia and Afghanistan, she could stir up sufficient trouble, possibly to completely overthrow British rule, but certainly to keep the English so occupied with uprisings as to force them to send troops to India rather than withdraw them thence for use elsewhere. The utter miscarriage of Germany's plans is, indeed, a fine tribute to Great Britain. The Emir of Afghanistan did probably more than any single native to thwart German treachery and intrigue, and every friend of the Allied cause must have read of his recent assassination with a very real regret.[10]

When General Maude took over the command, the effect of the Holy War that, at the Kaiser's instigation, was being preached in the mosques had not as yet been determined. This jehad, as it was called, proposed to unite all "True Believers" against the invading Christians, and give the war a strongly religious aspect. The Germans hoped by this means to spread mutiny among the Mohammedan troops, which formed such an appreciable element of the British forces, as well as to fire the fury of the Turks and win as many of the Arabs to their side as possible. The Arab thoroughly

10   Habibullah Khan (1872-1919) was emir and stayed neutral in World War One despite German and Turkish entreaties. He was assasinated in 1919, and his brother could not stay in power or keep the peace, as the Third Anglo-Afghan War ensued after.

disliked both sides. The Turk oppressed him, but did so in an Oriental, and hence more or less comprehensible, manner. The English gave him justice, but it was an Occidental justice that he couldn't at first understand or appreciate, and he was distinctly inclined to mistrust it. In course of time lie would come to realize its advantages. Under Turkish rule the Arab was oppressed by the Turk, but then he in turn could oppress the Jew, the Chaldean, and Nestorian Christians, and the wretched Armenian. Under British rule he suddenly found these latter on an equal footing with him, and he felt that this did not compensate the lifting from his shoulders of the Turkish burden. Then, too, when a race has been long oppressed and downtrodden, and suddenly finds itself on an equality with its oppressor, it is apt to become arrogant and overbearing. This is exactly what happened, and there was bad feeling on all sides in consequence. However, real fundamental justice is appreciated the world over, once the native has been educated up to it, and can trust in its continuity.

The complex nature of the problems facing the army commander can be readily seen. He was an indefatigable worker and an unsurpassed organizer. The only criticism I ever heard was that he attended too much to the details himself and did not take his subordinates sufficiently into his confidence. A brilliant leader, beloved by his troops, his loss was a severe blow to the Allied cause.

Baghdad is often referred to as the great example of the shattered illusion. We most of us have read the *Arabian Nights*[11] at an early age, and think of the abode of the caliphs as a dream city, steeped in what we have been brought up to think of as the luxury, romance, and glamour of the East. Now glamour is a delicate substance. In the all-searching glare of the Mesopotamian sun it is apt to appear merely tawdry. Still, a goodly number of years spent in wandering about in foreign lands had prepared me for a depreciation of the "stuff that dreams are made of,"[12] and I was not disappointed. It is unfortunate that the normal way to approach is from the south, and that that view of the city is flat and uninterest-

11   *The Thousand and One Nights* or *Arabian Nights,* translated into English by Richard Francis Burton.

12   *The Tempest,* Act IV, scene 1. "We are such stuff/As dreams are made on, and our little life/ Is rounded with a sleep."

ing. Coming, as I several times had occasion to, from the north, one first catches sight of great groves of date-palms, with the tall minarets of the Mosque of Kazimain towering above them; then a forest of minarets and blue domes, with here and there some graceful palm rising above the flat roofs of Baghdad. In the evening when the setting sun strikes the towers and the tiled roofs, and the harsh lights are softened, one is again in the land of Haroun-el-Raschid.[13]

The great covered bazaars are at all times capable of "eating the hours," as the natives say. One could sit indefinitely in a coffee-house and watch the throngs go by—the stalwart Kurdish porter with his impossible loads, the veiled women, the unveiled Christian or lower-class Arab women, the native police, the British Tommy, the kilted Scot, the desert Arab, all these and many more types wandered past. Then there was the gold and silver market, where the Jewish and Armenian artificers squatted beside their charcoal fires and haggled endlessly with their customers. These latter were almost entirely women, and they came both to buy and sell, bringing old bracelets and anklets, and probably spending the proceeds on something newer that had taken their fancy. The workmanship was almost invariably poor and rough. Most of the women had their babies with them, little mites decked out in cheap finery and with their eyelids thickly painted. The red dye from their caps streaked their faces, the flies settled on them at will, and they had never been washed. When one thought of the way one's own children were cared for, it seemed impossible that a sufficient number of these little ones could survive to carry on the race. The infant mortality must be great, though the children one sees look fat and thriving.

Baghdad is not an old city. Although there was probably a village on the site time out of mind, it does not come into any prominence until the eighth century of our era. As the residence of the Abasside caliphs it rapidly assumed an important position. The culmination of its magnificence was reached in the end of the eighth century, under the rule of the world-famous Haroun-el-Raschid. It long continued to be a centre of commerce and industry, though

---

13   Harun Al-Rashid (763-809) was the fifth Abbasid Caliph, considered to have overseen the beginning of the Islamic Golden Age.

suffering fearfully from the various sieges and conquests which it underwent. In 1258 the Mongols, under a grandson of the great Genghis Khan, captured the city and held it for a hundred years, until ousted by the Tartars under Tamberlane.[14] It was plundered in turn by one Mongol horde after another until the Turks, under Murad the Fourth, eventually secured it.[15] Naturally, after being the scene of so much looting and such massacres, there is little left of the original city of the caliphs. Then, too, in Mesopotamia there is practically no stone, and everything was built of brick, which readily lapses back to its original state.

For this reason the invaders easily razed a conquered town, and Mesopotamia, so often called the "cradle of the world," retains but little trace of the races and civilizations that have succeeded each other in ruling the land. When the Tigris was low at the end of the summer season, we used to dig out from its bank great bricks eighteen inches square, on which was still distinctly traced the seal of Nebuchadnezzar. These, possibly the remnants of a quay, were all that remained of the times before the advent of the caliphs.

---

14   Timur or Tamerlane (1320s-1405) was a Turkic-Mongol and founder of the Timurid dynasty. One of the last great nomadic conquerors following Genghis Khan.
15   Murad IV (1612-1640) Ottoman Sultan who waged war against the Abbasids and conquered Bagdhad after a siege in 1638. Fathered many children, wrote poetry. Died of cirrhosis at 27.

General Cobbe (center left facing camera), near Samarra, 1917

Chapter TWO
THE TIGRIS FRONT

A few days after reaching Baghdad I left for Samarra, which was at that time the Tigris front. I was attached to the Royal Engineers, and my immediate commander was Major Morin, D. S. O., an able officer with an enviable record in France and Mesopotamia. The advance army of the Tigris was the Third Indian Army Corps, under the command of General Cobbe, a possessor of the coveted, and invariably merited, Victoria Cross. The Engineers were efficiently commanded by General Swiney. The seventy miles of railroad from Baghdad to Samarra were built by the Germans, being the only Mesopotamian portion of the much-talked-of Berlin-to-Baghdad Railway, completed before the war. It was admirably constructed, with an excellent road-bed, heavy rails and steel cross-ties made by Krupp. In their retreat the Turks had been too hurried to accomplish much in the way of destruction other than burning down a few stations and blowing up the water-towers. The rolling-stock had been left largely intact. There were no passenger-coaches, and you travelled either by flat or box car. Every one followed the Indian custom of carrying with them their bedding-rolls, and leather- covered wash-basin containing their washing kit, as well as one of the comfortable rhoorkhee chairs. In consequence, although for travel by boat or train nothing was provided, there was no discomfort entailed. The trains were fitted out with anti-aircraft guns, for the Turkish aeroplanes occasionally tried to "lay eggs," a by no means easy affair with a moving train as a target. Whatever the reason was, and I never succeeded in discovering it, the trains invariably left Baghdad in the wee small hours, and as the station was on the right bank across the river from the main town, and the boat bridges were cut during the night, we used generally, when returning to the front, to spend the first part of the night sleeping on the station platform. Generals or exalted staff officers could usually succeed in having a car assigned to them, and hauled up from the yard in time for them to go straight to bed in it.

Frequently their trip was postponed, and an omniscient sergeant-major would indicate the car to the judiciously friendly, who could then enjoy a solid night's sleep. The run took anywhere from eight to twelve hours; but when sitting among the grain-bags on an open car, or comfortably ensconced in a chair in a "covered goods," with *Vingt Ans Apres*,[16] the time passed pleasantly enough in spite of the withering heat.While still a good number of miles away from Samarra we would catch sight of the sun glinting on the golden dome of the mosque, built over the cleft where the twelfth Imam, the Imam Mahdi, is supposed to have disappeared, and from which he is one day to reappear to establish the true faith upon earth. Many Arabs have appeared claiming to be the Mahdi, and caused trouble in a greater or less degree according to the extent of their following.[17] The most troublous one in our day was the man who besieged Kharthoum and captured General "Chinese" Gordon and his men. Twenty-five years later, when I passed through the Sudan, there were scarcely any men of middle age left, for they had been wiped out almost to a man under the fearful rule of the Mahdi, a rule which might have served as prototype to the Germans in Belgium.[18]

Samarra is very ancient, and has passed through periods of great depression and equally great expansion. It was here in A. D. 363 that the Roman Emperor Julian[19] died from wounds received in the defeat of his forces at Ctesiphon. The golden age lasted about forty years, beginning in 836, when the Caliph Hutasim[20] transferred his capital thither from Baghdad. During that time

---

16   Novel by Alexandre Dumas (1802-1870), during the French civil wars around the time of the English Civil War. A semi-sequel to the *Three Musketeers*.

17   The Mahdi is roughly equivalent to the Christan concept of the return of Christ for a final battle against Evil, as depicted in the Book of Revelations. Many leaders in Islamic history have claimed to be the Mahdi.

18   A strained comparison between the British war in Sudan, led by General Gordon, and the supposed German atrocities in Belgium, most of which were never proven to be true.

19   Julian the Apostate (331-361), last pagan Emperor of Rome.

20   al-Mu'tasim (796-842) is the eighth Abbasid Caliph. A very able ruler, administrator, patron and general. He founded Samarra.

the city extended for twenty-one miles along the river-bank, with glorious palaces, the ruins of some of which still stand. The present-day town has sadly shrunk from its former grandeur, but still has an impressive look with its great walls and massive gateways. The houses nearest the walls are in ruins or uninhabited; but in peacetime the great reputation that the climate of Samarra possesses for salubrity draws to it many Baghdad families who come to pass the summer months. A good percentage of the inhabitants are Persians, for the eleventh and twelfth Shiah Imanis are buried on the site of the largest mosque. The two main sects of Moslems are the Sunnis and the Shiahs; the former regard the three caliphs who followed Mohammed as his legitimate successors, whereas the latter hold them to be usurpers, and believe that his cousin and son-in-law, Ali, husband of Fatimah, together with their sons Husein and Hasan, are the prophet's true inheritors. Ali was assassinated near Nejef, which city is sacred to his memory, and his son Husein was killed at Kerbela; so these two cities are the greatest of the Shiah shrines. The Turks belong almost without exception to the Sunni sect, whereas the Persians and a large percentage of the Arabs inhabiting Mesopotamia are Shiahs.

The country around Samarra is not unlike in character the southern part of Arizona and northern Sonora. There are the same barren hills and the same glaring heat. The soil is not sand, but a fine dust which permeates everything, even the steel uniform-cases which I had always regarded as proof against all conditions. The parching effect was so great that it was not only necessary to keep all leather objects thoroughly oiled but the covers of my books cracked and curled up until I hit upon the plan of greasing them well also. In the alluvial lowlands trench-digging was a simple affair, but along the hills we found a pebbly conglomerate that gave much trouble.

The War In

Golden Dome of Samarra

Opinion was divided as to whether the Turk would attempt to advance down the Tigris. Things had gone badly with our forces in Palestine at the first battle of Gaza; but here we had an exceedingly strong position, and the consensus of opinion seemed to be that the enemy would think twice before he stormed it., Their base was at Tekrit, almost thirty miles away. However, about ten miles distant stood a small village called Daur, which the Turks held in considerable force. Between Daur and Samarra there was nothing but desert, with gazelles and jackals the only permanent inhabitants. Into this no man's land both sides sent patrols, who met in occasional skirmishes. For reconnaissance work we used light-armored motor-cars, known throughout the army as Lam cars, a name formed by the initial letters of their titles. These cars were Rolls-Royces, and with their armor-plate weighed between three and three-quarters and four tons. They were proof against the ordinary bullet but not against the armor-piercing. When I came out to Mesopotamia I intended to lay my plans for a transfer to the cavalry, but after I had seen the cars at work I changed about and asked to be seconded to that branch of the service.

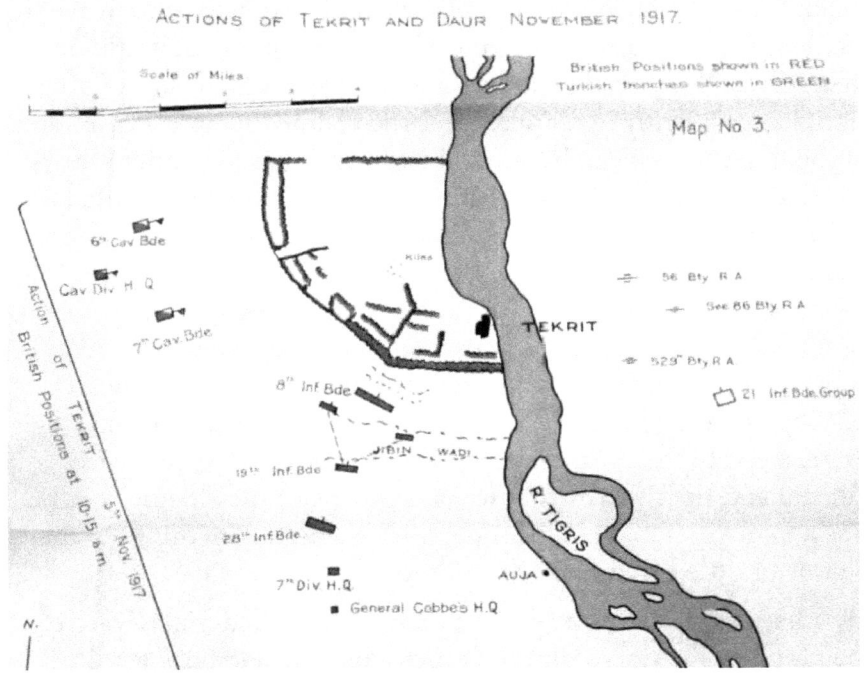

A short while after my arrival our aeroplanes brought in word that the Turks were massing at Daur, and General Cobbe decided that when they launched forth he would go and meet them. Accordingly, we all moved out one night, expecting to give "Abdul," as the Tommies called him, a surprise. Whether it was that we started too early and their aero-planes saw us, or whether they were only making a feint, we never found out; but at all events the enemy fell back, and save for some advance-guard skirmishing and a few prisoners, we drew a blank. We were not prepared to attack the Daur position, and so returned to Samarra to await developments.

Meanwhile I busied myself searching for an Arab servant. Seven or eight years previous, when with my father in Africa, I had learned Swahili, and although I had forgotten a great deal of it, still I found it a help in taking up Arabic. Most of the officers had either British or Indian servants; in the former case they were known as batmen, and in the latter as bearers; but I decided to follow suit with the minority and get an Arab, and therefore learn Arabic instead of Hindustanee, for the former would be of vastly more general use. The town commandant, Captain Grieve of the Black Watch, after many attempts at length produced a native who seemed, at any rate, more promising than the others that offered themselves. Yusuf was a sturdy, rather surly-looking youth of about eighteen. Evidently not a pure Arab, he claimed various admixtures as the fancy took him, the general preference being Kurd. I always felt that there was almost certainly a good percentage of Turk. His father had been a non-commissioned officer in the Turkish army, and at first I was loath to take him along on advances and attacks, for he would have been shown little mercy had he fallen into enemy hands. He was, however, insistent on asking to go with me, and I never saw him show any concern under fire. He spoke, in varying degrees of fluency, Kurdish, Persian, and Turkish, and was of great use to me for that reason. He became by degrees a very faithful and trustworthy follower, his great weakness being that he was a one-man's man, and although he would do anything for me, he was of little general use in an officers' mess.

I had two horses, one a black mare that I called Soda, which means black in Arabic, and the other a hard-headed bay gelding that was game to go all day, totally unaffected by shell-fire, but exceedingly stubborn about choosing the direction in which he went. After numerous changes I came across an excellent syce to look after them. He was a wild, unkempt figure, with a long black beard—a dervish by profession, and certainly gave no one any reason to believe that he was more than half-witted. Indeed, almost all dervishes are in a greater or less degree insane; it is probably due to that that they have become dervishes, for the native regards the insane as under the protection of God.[21] Dervishes go around practically naked, usually wearing only a few skins flung over the shoulder, and carrying a large begging-bowl. In addition they carry a long, sharp, iron bodkin, with a wooden ball at the end, having very much the appearance of a fool's bauble. They lead an easy life. When they take a fancy to a house, they settle down near the gate, and the owner has to support them as long as the whim takes them to stay there. To use force against a dervish would be looked upon as an exceedingly unpropitious affair to the true believer. Then, too, I have little doubt but that they are capable of making good use of their steel bodkin. Why my dervish wished to give up his easy-going profession and take over the charge of my horses I never fully determined, but it must have been because he really loved horses and found that as a dervish pure and simple he had very little to do with them. When he arrived he was dressed in a very ancient gunny-sack, and it was not without much regret at the desecration that I provided him with an outfit of the regulation khaki.

My duties took me on long rides about the country. Here, and throughout Mesopotamia, the great antiquity of this "cradle of the world" kept ever impressing itself upon one, consciously or subconsciously. Everywhere Were ruins; occasionally a wall still reared itself clear of the all-enveloping dust, but generally all that remained were great mounds, where the desert had crept in and claimed its own, covering palace, house, and market, temple, synagogue, mosque, or church with its everlasting mantle. Often the streets could still be traced, but oftener not. The weight of ages was ever present as one rode among the ruins of these once busy, prosperous cities, now long dead and buried, how long no one knew,

21   A related concept to the whirling dervish is the Russian Orthodox notion of 'yurodivy,' or Holy Fool.

for frequently their very names were forgotten. Babylon, Ur of the Chaldees, Istabulat, Nineveh, and many more great cities of history are now nothing but names given to desert mounds.

Close by Samarra stands a strange corkscrew tower, known by the natives as the Malwiyah. It is about a hundred and sixty feet high, built of brick, with a path of varying width winding up around the outside. No one knew its purpose, and estimates of its antiquity varied by several thousand years. One fairly well-substantiated story told that it had been the custom to kill prisoners by hurling them off its top. We found it exceedingly useful as an observation-post. In the same manner we used Julian's tomb, a great mound rising up in the desert some five or six miles up-stream of the town. The legend is that when the Roman Emperor died of his wounds his soldiers, impressing the natives, built this as a mausoleum; but there is no ground whatever for this belief, for it would have been physically impossible for a harassed or retreating army to have performed a task of such magnitude. The natives call it "The Granary," and claim that that was its original use. Before the war the Germans had started in excavating, and discovered shafts leading deep down, and on top the foundations of a palace. Around its foot may be traced roadways and circular plots, and especially when seen from an aero plane it looks as if there had at one time been an elaborate system of gardens.

We were continually getting false rumors about the movements of the Turks. We had believed that it would be impossible for them to execute a flank movement, at any rate in sufficient strength to be a serious menace, for from all the reports we could get, the wells were few and far between. Nevertheless, there was a great deal of excitement and some concern when one afternoon our aeroplanes came in with the report that they had seen a body of Turks that they estimated at from six to eight thousand marching round our right flank. The plane was sent straight back with instructions to verify most carefully the statement, and be sure that it was really men they had seen. They returned at dark with no alteration of their original report. As can well be imagined, that night was a crowded one for us, and the feeling ran high when next morning the enemy turned out to be several enormous herds of sheep.

As part consequence of this we were ordered to make a thorough water reconnaissance, with a view of ascertaining how large a force could be watered on a march around our flank. I went off in an armored car with Captain Marshall of the Intelligence Service. Marshall had spent many years in Mesopotamia shipping liquorice to the American Tobacco Company, and he was known and trusted by the Arabs all along the Tigris from Kurna to Mosul. He spoke the language most fluently, but with an accent that left no doubt of his Caledonian home. We had with us a couple of old sheiks, and it was their first ride in an automobile. It was easy to see that one of them was having difficulty in maintaining his dignity, but I was not quite sure of the reason until we stopped a moment and he fairly flew out of the car. It didn't seem possible that a man able to ride ninety miles at a stretch on a camel, could be made ill by the motion of an automobile. However, such was the case, and we had great difficulty in getting him back into the car. We discovered far more wells than we had been led to believe existed, but not enough to make a flank attack a very serious menace.

The mirage played all sorts of tricks, and the balloon observers grew to be very cautious in their assertions. In the early days of the campaign, at the battle of Shaiba Bund, a friendly mirage saved the British forces from what would have proved a very serious defeat. Suleiman Askari was commanding the Turkish forces, and things were faring badly with the British, when of a sudden to their amazement they found that the Turks were in full retreat. Their commanders had caught sight of the mirage of what was merely an ambulance and supply train, but it was so magnified that they believed it to be a very large body of reinforcements. The report ran that when Suleiman was told of his mistake, his chagrin was so great that he committed suicide.

It was at length decided to advance on the Turkish forces at Daur. General Brooking had just made a most successful attack on the Euphrates front, capturing the town of Ramadie, with almost five thousand prisoners. It was believed to be the intention of the army commander to try to relieve the pressure against' General Allenby's forces in Palestine by attacking the enemy on all three of their Mesopotamian fronts. Accordingly, we were ordered to march out after sunset one night, prepared to attack the enemy position at day-break. During a short halt by the last rays of the setting sun I caught sight of a number of Mohammedan soldiers prostrating themselves toward Mecca in their evening prayers, while their Christian or pagan comrades looked stolidly on. It was late October, and although the days were still very hot and oppressive, the nights were almost bitterly cold. A night march is always a disagreeable business. The head of the column checks and halts, and those in the rear have no idea whether it is an involuntary stop for a few minutes, or whether they are to halt for an hour or more, owing to some complication of orders. So we stood shivering, and longed for a smoke, but of course that was strictly forbidden, for the cigarettes of an army would form a very good indication of its whereabouts on a dark night. All night we marched and halted, and started on again; the dust choked us, and the hours seemed in-terminable, until at last at two in the morning word was passed along that we could have an hour's sleep. The greater part of the year in Mesopotamia the regulation army dress consisted of a tunic and "shorts." These are long trousers cut off just above the knee, and the wearer may either use wrap puttees, or leather leggings, or golf stockings. They are a great help in the heat, as may easily be understood, and they allow, of course, much freer knee action, particularly when your clothes are wet. The reverse side of the medal reads that when you try to sleep without a blanket on a cold night, you find that your knees are uncomfortably exposed. Still we were, most of us, so drunk with sleep that it would have taken more than that to keep us awake. At three we resumed our march, and attacked just at dawn. The enemy had abandoned the first- line positions, and we met with but little resistance in the second. Our cavalry, which was concentrated at several points in nullahs (dry river-beds), suffered at the hands of the hostile aircraft. The Turk had evidently determined to fall back to Tekrit without putting up a serious defense. They certainly could have given us a much worse time than they did, for they had dug in well and scientifically. Among the prisoners we took there were some that proved to be very worth while. These Turkish officers were, as a whole a good lot—well dressed and well educated. Many spoke French. There is an excellent gunnery school at Constantinople, and one of the officers we captured had been a senior instructor there for many years. We had with us among our intelligence officers a Captain Bettelheim, born in Constantinople of Belgian parentage. He had served with the Turks against the Italians and with the British against the Boers. This gunnery officer turned out to be an old comrade of his in the Italian War.[22] Many of the officers we got knew him, for he had been chief of police in Constantinople. Apparently none of them bore him the slightest ill-will when they found him serving against them.

Among the supplies we captured at Daur were a lot of our own rifles and ammunition that the Arabs had stolen and sold to the Turks. It was impossible to entirely stop this, guard our dumps as best we could. On dark nights they would creep right into camp, and it was never safe to have the hospital barges tie up to the banks for the night on their way down the river. On many occasions the Arabs crawled aboard and finished off the wounded. There was only one thing to be said for the Arab, and that was that he played no favorite, but attacked, as a rule, whichever side came handier. We were told, and I believe it to be true, that during the fighting at Sunnaiyat the Turks sent over to know if we would agree to a three days' truce, during which time we should join forces against the Arabs, who were watching on the flank to pick off stragglers or ration convoys.

That night we bivouacked at Daur, and were unmolested except for the enemy aircraft that came over and "laid eggs." Next morning we advanced on Tekrit. Our orders were to make a feint, and if we found that the Turk meant to stay and fight it out seriously, we were to fall back. Some gazelles got into the no man's land between us and the Turk, and in the midst of the firing ran gracefully up the line, stopping every now and then to stare about in amazement. Later on in the Argonne forest in France we had the same thing happen with some wild boars. The enemy seemed in no way inclined to evacuate Tekrit, so in accordance with instructions we returned to our previous night's encampment at Daur. On the way

---

22   The Italian-Turkish War, from 1911 to 1912, when Italy annexed the Dodecanese and Libya from the Ottomans.

back we passed an old "arabana," a Turkish coupe, standing abandoned in the desert, with a couple of dead horses by it. It may have been used by some Turkish general in the retreat of two days before. It was the sort of coupe one associates entirely with wellkept parks and crowded city streets, and the incongruity of its lonely isolation amid the sand-dunes caused an amused ripple of comment.

Our instructions were to march back to Samarra early next morning, but shortly before midnight orders came through from General Maude for us to advance again upon Tekrit and take it. Next day we halted and took stock in view of the new orders. The cavalry again suffered at the hands of the Turkish aircraft. I went to corps headquarters in the afternoon, and a crowd of "red tabs," as the staff-officers were called, were seated around a little table having the inevitable tea. A number of the generals had come in to discuss the plan of attack for the following day. Suddenly a Turk aeroplane made its appearance, flying quite low, and dropping bombs at regular intervals. It dropped two, and then a third on a little hill in a straight line from the staff conclave. It looked as if the next would be a direct hit, and the staff did the only wise thing, and took cover as flat on the ground as nature would allow; but the Hun's spacing was bad, and the next bomb fell some little way beyond. I remember our glee at what we regarded as a capital joke on the staff. The line-officer's humor becomes a trifle robust where the "gilded staff" is concerned, notwithstanding the fact that most staff-officers have seen active and distinguished service in the line.

Our anti-aircraft guns—"Archies" we called them—were mounted on trucks, and on account of their weight had some difficulty getting up. I shall not soon forget our delight when they lumbered into view, for although I never happened personally to see an aeroplane brought down by an "Archie," there was no doubt about it but that they did not bomb us with the same equanimity when our anti-aircrafts were at hand.

That night we marched out on Tekrit, and as dawn was breaking were ready to attack. As the mist cleared, an alarming but ludicrous sight met our eyes. On the extreme right some caterpillar tractors hauling our "heavies" were advancing straight on Tekrit, as if they had taken themselves for tanks. They were not long in discovering their mistake, and amid a mixed salvo they clumsily turned and made off at their best pace, which was not more than three miles an hour. Luckily, they soon got under some excellent defilade, but not until they had suffered heavily.

British and Australian Camel Troops.

Captured Turkish camel corps

Our artillery did some good work, but while we were waiting to attack we suffered rather heavily. We had to advance over a wide stretch of open country to reach the Turkish first lines. By nightfall the second line of trenches was practically all in our hands. Meanwhile the cavalry had circled way around the flank up-stream of Tekrit to cut the enemy off if he attempted to retreat. The town is on the right bank of the Tigris, and we had a small force that had come up from Samarra on the left bank, for we had no means of ferrying troops across. Our casualties during the day had amounted to about two thousand. The Seaforths had suffered heavily, but no more so than some of the native regiments. In Mesopotamia there were many changes in the standing of the Indian battalions. The Maharattas, for instance, had never previously been regarded as anything at all unusual, but they have now a very distinguished record to take pride in. The general feeling was that the Gurkhas did not quite live up to their reputation. But the Indian troops as a whole did so exceedingly well that there is little purpose in making comparisons amongst them. At this time, so I was informed, the Expeditionary Force, counting all branches, totalled about a million, and a very large percentage of this came from India. We drew our supplies from India and Australia, and it is interesting to note that we preferred the Australian canned beef and mutton (bully beef and bully mutton, as it was called) to the American.

At dusk the fighting died down, and we were told to hold on and go over at daybreak. As I was making my way back to headquarters a general pounced upon me and told me to get quickly into a car and go as rapidly as possible to Daur to bring up a motor ration convoy with fodder for the cavalry horses and food for the riders. A Ford car happened to pass by, and he stopped it and shoved me in, with some last hurried injunction. It was quite fifteen miles back, and the country was so cut up by nullahs or ravines that in most places it was inadvisable to leave the road, which was, of course, jammed with a double stream of transport of every description. When we were three or four miles from Daur a tire blew out. The driver had used his last spare, so there was nothing to do but keep going on the rim. The car was of the delivery-wagon type—"pill-boxes" were what they were known as—and while we were stopped taking stock I happened to catch sight of a good-sized bedding-roll behind. "Some one's out of luck," said I to the driver; "whose roll is it?" "The corps commander's, sir," was his reply. After exhausting my limited vocabulary, I realized that it was far too late to stop another motor and send this one back, so I just kept going. Across the bed of one more ravine, the sand up to the hubs, and we were in the Daur camp. I managed to rank some one out of a spare tire and started back again. My driver proved unable to drive at night, at all events at a pace that would put us anywhere before dawn, so I was forced to take the wheel. By the time I had the convoy properly located I was rather despondent of the corps commander's temper, even should I eventually reach him that night, which seemed a remote chance, for the best any one could do was give me the rough location on a map. Still, taking my luminous compass, I set out to steer a cross-country course. I ran into five or six small groups of ambulances filled with wounded, trying to find their way to Daur, and completely lost. Most had given up— some were unknowingly headed back for Tekrit. I could do no more than give them the right direction, which I knew they had no chance of holding. Of course I could have no headlights, and the ditches were many, but in some miraculous way, more through good luck than good management, I did find corps headquarters, and what was better still, the general's reprimand took the form of bread and ham and a stiff peg of whiskey—the first food I had had since before daylight.

During the night the Turks evacuated the town. Their forces were certainly mobile. They could cover the most surprising distances, and live on almost nothing. We marched in and occupied. White flags were flying from all the houses, which were not nearly so much damaged from the bombardment as one would have supposed. This was invariably the case; indeed, it is surprising to see how much shelling a town can undergo without noticeable effect. It takes a long time to level a town in the way it has been done in northern France. In this region the banks of the river average about one hundred and fifty feet in height, and Tekrit is built at the junction of two ravines. No two streets are on the same level; sometimes the roofs of the houses on a lower level serve as the streets for the houses above. Many of the booths in the bazaar were open and transacting business when we arrived, an excellent proof of how firmly the Arabs believed in British fair dealing. Our men bought cigarettes, matches, and vegetables. Yusuf had lived here three or four years, so I despatched him to get chickens and eggs for the mess. I ran into Marshall, who was on his way to dine with the mayor, who had turned out to be an old friend of his. He asked me to join him, and we climbed up to a very comfortable house, built around a large courtyard. It was the best meal we had either of us had in days —great pilaus of rice, excellent chicken, and fresh unleavened bread. This bread looks like a very large and thin griddle-cake. The Arab uses it as a plate. Eating with your hands is at first rather difficult. Before falling to, a ewer is brought around to you, and you are supplied with soap—a servant pours water from the ewer over your hands, and then gives you a towel. After eating, the same process is gone through with. There are certain formalities that must be regarded—one of them being that you must not cat or drink with your left hand.

In Tekrit we did not find as much in the way of supplies and ammunition as we had hoped. The Turk had destroyed the greater part of his store. We did find great quantities of wood, and in that barren, treeless country it was worth a lot. Most of the inhabitants of Tekrit are raftsmen by profession. Their rafts have been made in the same manner since before the days of Xerxes and Darius. Inflated goatskins are used as a basis for a platform of poles, cut in the up-stream forests. On these, starting from Diarbekr or Mosul, they float down all their goods. When they reach Tekrit they leave the poles there, and start up-stream on foot, carrying their deflated goatskins. The Turks used this method a great deal bringing down their supplies. In pre-war days the rafts, keleks as they are called, would often come straight through to Baghdad, but many were always broken up at Tekrit, for there is a desert route running across to Hit on the Euphrates, and the supplies from up-river were taken across this in camel caravans.

The aerodrome lay six or seven miles above the town, and I was anxious to see it and the comfortable billets the Germans had built themselves. I found a friend whose duties required motor transportation, and we set off in his car. A dust-storm was raging, and we had some difficulty in finding our way through the network of trenches. Once outside, the storm became worse, and we could only see a few yards in front of us. We got completely lost, and after nearly running over the edge of the bluff, gave up the attempt,' and slowly worked our way back.

When we started off on the advance I was reading Xenophon's _Anabasis_.[23] On the day when we were ordered to march on Tekrit

23   The Anabasis is a famous narrative about a Greek mercenary army,

a captain of the Royal Flying Corps, an ex-master at Eton, was in the mess, and when I told him that I was nearly out of reading matter, he said that next time he came over he would drop me Plutarch's *Lives*.[24] I asked him to drop it at corps headquarters, and that a friend of mine there would see that I got it. The next day in the heat of the fighting a plane came over low, signalling that it was dropping a message. As the streamer fell close by, there was a rush to pick it up and learn how the attack was progressing. Fortunately, I was far away when the packet was opened and found to contain the book that the pilot had promised to drop for me.

After we had been occupying the town for a few days, orders came through to prepare to fall back on Samarra. The line of communication was so long that it was impossible to maintain us, except at too great a cost to the transportation facilities possessed by the Expeditionary Forces. Eight or ten months later, when we had more rails in hand, a line was laid to Tekrit, which had been abandoned by the Turks under the threat of our advance to Kirkuk, in the Persian hills. It was difficult to explain to the men, particularly to the Indians, the necessity for falling back. All they could understand was that we had taken the town at no small cost, and now we were about to give it up.

For several days I was busy helping to prepare rafts to take down the timber and such other captured supplies as were worth removing. The river was low, leaving a broad stretch of beach below the town, and to this we brought down the poles. Several camels had died near the water, probably from the results of our shelling, and the hot weather soon made them very unpleasant companions. The first day was bad enough; the second was worse. The natives were not in the least affected. They brought their washing and worked among them—they came down and drew their drinking-water from the river, either beside the camels or down-stream of them, with complete indifference. It is true this water percolates drop by drop through large, porous clay pots before it is drunk, but even so, it would have seemed that they would have preferred its

the Ten Thousand, who were recruited to help Cyrus the Younger claim the Persian throne. He died, and they had to fight their way home.

24   Plutarch's *Lives* is a compilation of Greek and Roman biographical sketches of great men. One of the greatest masterpieces of the Western Canon.

coming from up-stream of the derelict "ships of the desert." On the third day, to their mild surprise, we managed with infinite difficulty to tow the camels out through the shallow water into the main stream.

We finally got our rafts built, over eighty in number, and arranged for enough Arab pilots to take care of half of them. On the remainder we put Indian sepoys. They made quite a fleet when we finally got them all started down-stream. Two were broken up in the rapids near Daur, the rest reached Samarra in safety on the second day.

We had a pleasant camp on the bluffs below Tekrit—high enough above the plain to be free of the ordinary dust-storms, and the prospect of returning to Samarra was scarcely more pleasant to us than to the men. Five days after we had taken the town, we turned our backs on it and marched slowly back to rail-head.

CHAPTER THREE
PATROLLING THE RUINS OF BABYLON

We returned to find Samarra buried in dust and more desolate than ever. A few days later came the first rain-storm. After a night's downpour the air was radiantly clear, and it was joy to ride off on the rounds, no longer like Zeus, enveloped in a cloud.

It was a relief to see the heat-stroke camps broken up. During the summer months our ranks were fearfully thinned through the sun. Although it was the British troops that suffered most, the Indians were by no means immune. Before the camps were properly organized the percentage of mortality was exceedingly large, for the only effective treatment necessitates the use of much ice. The patient runs a temperature which it was impossible to control until the ice-making machines were installed. The camps were situated in the coolest and most comfortable places, but in spite of everything, death was a frequent result, and recoveries were apt to be only partial. Men who had had a bad stroke were rarely of any further use in the country.

Another sickness of the hot season which now began to claim less victims was sandfly fever. This fever, which, as its name indicates, was contracted from the bites of sandflies, varied widely in virulence. Sometimes it was so severe that the victim had to be evacuated to India; as a rule he went no farther than a base hospital at Baghdad or Amara.

One of the things about which the Tommy felt most keenly in the Mesopotamian campaign was that there was no such thing as a "Cushy Blighty." To take you to "Blighty" a wound must mean permanent disablement, otherwise you either convalesced in the country or, at best, were sent to India. In the same manner there were no short leaves, for there was nowhere to go. At the most rapid rate of travelling it took two weeks to get to India, and once there, although the people did everything possible in the way of entertaining, the enlisted man found little to make him less homesick than he had been in Mesopotamia. Transportation was so difficult and the trip so long that only under very exceptional circumstances was leave to England given. One spring it was announced that officers wishing to get either married or divorced could apply for leave with good hopes of success. Many applied, but a number returned without having fulfilled either condition, so that the following year no leaves were given upon those grounds. The army commander put all divorce cases into the hands of an officer whose civil occupation had been the law, and who arranged them without the necessity of granting home leave.

A week after our return to Samarra a rumor started that General Maude was down with cholera. For some time past there had been sporadic cases, though not enough to be counted an epidemic. The sepoys had suffered chiefly, but not exclusively, for the British ranks also supplied a quota of victims. An officer on the staff of the military governor of Baghdad had recently died. We heard that the army commander had the virulent form, and knew there could be no chance of his recovery. The announcement of his death was a heavy blow to all, and many were the gloomy forebodings. The whole army had implicit confidence in their leader, and deeply mourned his loss. The usual rumors of foul play and poison went the rounds, but I soon after heard Colonel Wilcox—in pre-war days an able and renowned practitioner of Harley Street—say that it was an undoubted case of cholera. The colonel had attended General Maude throughout the illness. The general had never taken the cholera prophylactic, although Colonel Wilcox had on many occasions urged him to do so, the last time being only a few days before the disease developed.

General Marshall, who had commanded General Maude's old division, the Thirteenth, took over. The Seventeenth lost General Gillman, who thereupon became chief of staff. This was a great loss to his division, for he was the idol of the men, but the interest of the Expeditionary Force was naturally and justly given precedence.

In due course my transfer to the Motor Machine-Gun Corps

came through approved, and I was assigned to the Fourteenth battery of light-armored motor-cars, commanded by Captain Nigel Somerset, whose grandfather, Lord Raglan, had died, nursed by Florence Nightingale, while in command of the British forces in the Crimean War. Somerset himself was in the infantry at the outbreak of the war and had been twice wounded in France. He was an excellent leader, possessing as he did dash, judgment, and personal magnetism. A battery was composed of eight armored cars, subdivided into lour sections. There was a continually varying number of tenders and workshop lorries. The fighting cars were Rolls-Royces, the others Napiers and Fords.

Below: original pictures of armored cars. Opposite: more armored Rolls Royces in the Mesopotamian front.

Towing an armored car across a river

Reconnaissance

At that time there were only four batteries in the country. We were army troops—that is to say, we were not attached to any individual brigade, or division, or corps, but were temporarily assigned first here and then there, as the need arose.

In attacks we worked in co-operation with the cavalry. Although on occasions they tried to use us as tanks, it was not successful, for our armor-plate was too light. We were also employed in raiding, and in quelling Arab uprisings. This latter use threw us into close touch with the political officers. These were a most interesting lot of men. They were recruited in part from the army, but largely from civil life. They took over the civil administration of the conquered territory and judiciously upheld native justice. Many remarkable characters were numbered among them—men who had devoted a lifetime to the study of the intricacies of Oriental diplomacy. They were distinguished by the white tabs on the collars of their regulation uniforms; but white was by no means invariably the sign of peace, for many of the political officers were killed, and more than once in isolated towns in unsettled districts they sustained sieges that lasted for several days. We often took a political officer out with us on a raid or reconnaissance, finding his knowledge of the language and customs of great assistance. Sir Percy Cox was at the head, with the title "Chief Political Officer" and the rank of general. His career in the Persian Gulf has been as distinguished as it is long, and his handling of the very delicate situations arising in Mesopotamia has called forth the unstinted praise of soldier and civilian alike.

Ably assisting him, and head of the Arab bureau, was Miss Gertrude Bell, the only woman, other than the nursing sisters, officially connected with the Mesopotamian Expeditionary Forces. Miss Bell speaks Arabic fluently and correctly. She first became interested in the East when visiting her uncle at Teheran, where he was British minister. She has made noteworthy expeditions in Syria and Mesopotamia, and has written a number of admirable books, among which are *Armurath to Armurath* and *The Desert and the Sown*. [25]The undeniable position which she holds must appear

---

25   Gertrude Bell, CBE (1868-1926) was born to a wealthy family, and became a well known Arabist and archeologist. She worked alongside T.E.

doubly remarkable when the Mohammedan official attitude toward women is borne in mind. Miss Bell has worked steadily and without a leave in this trying climate, and her tact and judgment have contributed to the British success to a degree that can scarcely be overestimated.

The headquarters of the various batteries were in Baghdad. There we had our permanent billets, and stores. We would often be ordered out in sections to be away varying lengths of time, though rarely more than a couple of months. The workshops' officer stayed in permanent charge and had the difficult task of keeping all the cars in repair. The supply of spare parts was so uncertain that much skill and ingenuity were called for, and possessed to a full degree by Lieutenant Linnell of the Fourteenth.

A few days after I joined I set off with Somerset and one of the battery officers, Lieutenant Smith, formerly of the Black Watch. We were ordered to do some patrolling near the ruins of Babylon. Kerbela and Nejef, in the quality of great Shiah shrines, had never been particularly friendly to the Turks, who were Sunnis—but the desert tribes are almost invariably Sunnis, and this coupled with their natural instinct for raiding and plundering made them eager to take advantage of any interregnum of authority. We organized a sort of native mounted police, but they were more picturesque than effective. They were armed with weapons of varying age and origin—not one was more recent than the middle of the last century. Now the Budus, the wild desert folk, were frequently equipped with rifles they had stolen from us, so in a contest the odds were anything but even.

We took up our quarters at Museyib, a small town on the banks of the Euphrates, six or eight miles above the Hindiyah Barrage, a dam finished a few years before, and designed to irrigate a large tract of potentially rich country. We patrolled out to Mohamediyah, a village on the caravan desert route to Baghdad, and thence down to Hilleh, around which stand the ruins of ancient Babylon.

Lawrence to lobby the British government to set up independent Arab states in the Mid-East. She advised King Faisal and Wisnton Churchill. She wrote many books. She was appointed head of antiquities by Faisal and founded the Baghdad Archeological Museum.

The rainy season was just beginning, and it was obvious that the patrolling could not be continuous, for a twelve-hour rain would make the country impassable to our heavy cars for two or three days. We were fortunate in having pleasant company in the officers of a Punjabi infantry battalion and an Indian cavalry regiment. Having commandeered an ancient caravan-serai for garage and billets, we set to work to clean it out and make it as waterproof as circumstances would permit. An oil-drum with a length of iron telegraph-pole stuck in its top provided a serviceable stove, and when it rained we played bridge or read.

I was ever ready to reduce my kit to any extent in order to have space for some books, and Voltaire's *Charles XII* was the first called upon to carry me to another part of the world from that in which I at the moment found myself.[26] I always kept a volume of some sort in my pocket, and during halts I would read in the shade cast by the turret of my car. The two volumes of Layard's *Early Adventures* proved a great success. The writer, the great Assyriologist, is better known as the author of *Nineveh and Babylon*.[27] The book I was reading had been written when he was in his early twenties, but published for the first time forty years later. Layard started life as a solicitor's clerk in London, but upon being offered a post in India he had accepted and proceeded thither overland. On reaching Baghdad he made a side-trip into Kurdistan, and became so enamored of the life of the tribesmen that he lived there with them on and off for two years—years filled with adventure of the most thrilling sort.

I had finished a translation of Xenophon shortly before and found it a very different book than when I was plodding drearily through it in the original at school. Here it was all vivid and real before my eyes, with the scene of the great battle of Cunaxa only a few miles from Museyib. Babylon was in sight of the valiant

---

26   Voltaire wrote a biography of Charles XII (1682-1715) or Carolus Rex, the King of Sweden, in 1731. He was a masterful general, defeating the Russians twice at Narva in 1700 and Holowczyn in 1708.
27   Austen Henry Layard (1817-1894) was an English Assyriologist, historion, politician and diplomat. He served in the Russell, Palmerston, and Gladstone cabinets in the 1860s, and as ambassador to the Ottomans from 1877-1880.

Greeks, but all through the loss of a leader it was never to be theirs. On the ground itself one could appreciate how great a master-piece the retreat really was, and the hardiness of the soldiers which caused Xenophon to regard as a "snow sickness" the starvation and utter weariness which made the numbed men lie down and die in the snow of the Anatolian highlands. He remarks naively that if you could build a fire and give them something hot to eat, the sickness was dispelled!

The rain continued to fall and the mud became deeper and deeper. It was all the Arabs could do to get their produce into market. The bazaar was not large, but was always thronged. I used to sit in one of the coffee-houses and drink coffee or tea and smoke the long-stemmed water-pipe, the narghile. My Arabic was now sufficiently fluent for ordinary conversation, and in these clubs of the Arab I could hear all the gossip. Bazaar rumors always told of our advances long before they were officially given out. Once in Baghdad I heard of an attack we had launched. On going around to G.H.Q. I mentioned the rumor, and found that it was not yet known there, but shortly after was confirmed. I had already in Africa met with the "native wireless," and it will be remembered how in the Civil War the plantation negroes were often the first to get news of the battles. It is something that I have never heard satisfactorily explained.

In the coffee-houses, besides smoking and gossiping, we also played games, either chess or backgammon or munkula. This last is an exceedingly primitive and ancient game—it must date almost as far back as jackstones or knucklebones. I have seen the natives in Central Africa and the Indians in the far interior of Brazil playing it in almost identical form. In Mesopotamia the board was a log of wood sliced in two and hinged together. In either half five or six holes were scooped out, and the game consisted in dropping cowrie shells or pebbles into the holes. When the number in a particular hollow came to a certain amount with the addition of the one dropped in, you won the contents.

In most places the coffee was served in Arab fashion, not Turkish. In the latter case it is sweet and thick and the tiny cup is half full of grounds; in the former the coffee is clear and bitter and of unsurpassable flavor. The diminutive cup is filled several times, but each time there is only a mouthful poured in. Tea is served in small glasses, without milk, but with lots of sugar. The spoons in the glasses are pierced with holes like tea-strainers so that the tea may be stirred without spilling it.

There was in particular one booth I could never tire watching. The old man who owned it was a vender of pickles. In rows before him were bottles and jars and bowls containing pickles of all colors—red, yellow, green, purple, white, and even blue. Above his head were festoons of gayly painted peppers. He had a long gray beard, wore a green turban and a flowing robe with a gold-braided waistcoat. In the half-lights of the crowded, covered bazaar his was a setting in which Dulac would have revelled.[28]

At Museyib we led a peaceful, uneventful existence—completely shut in by the mud. We had several bazaar rumors about proposed attacks upon the engineers who were surveying for a railroad that was to be built to Hilleh for the purpose of transporting the grain-crop to the capital. Nothing materialized, however. The conditions were too poor to induce even the easily encouraged Arabs to raid. One morning when I was wandering around the gardens on the outskirts of the town I came across some jackals and shot one with my Webley revolver. It was running and I fired a number of times, and got back to town to find that my shooting had started all sorts of excitement and reports of uprisings.

Christmas came and the different officers' messes organized celebrations. The mess we had joined was largely Scotch, so we decided we must make a haggis, that "chieftain of the pudden race." The ingredients, save for the whiskey, were scarcely orthodox, but if it was not a success, at least no one admitted it.

As soon as the weather cleared we made a run to Kerbela—a lovely town, with miles of gardens surrounding it and two great mosques. The bazaar was particularly attractive—plentifully supplied with everything. We got quantities of the deliciously flavored

28   Edmund Dulac (1882-1953) was an illustrator and stamp designer. He designed banknotes for the UK during the Second World War. His children's and classics books illustrations are what Kermit Roosevelt has in mind.

pistachio-nuts which were difficult to obtain elsewhere, as well as all sorts of fruit and vegetables. There were no troops stationed in the vicinity, so the prices were lower than usual. The orders were that we should go about in armed bands, but I never saw any marked indication of hostility. The British, true to the remarkable tact and tolerance that contributes so largely to their success in dealing with native races, posted Mohammedan sepoys as guards on the mosques, and no one but Moslems could even go into the courtyards. If this had not been done, there would have been many disturbances and uprisings, for the Arabs and Persians felt so strongly on the question that they regarded with marked hostility those who even gazed into the mosque courtyards. Why it is so different in Constantinople I do not know, but there was certainly no hostility shown us in Santa Sophia nor in the mosque of Omar in Jerusalem. Be that as it may, forbidden fruit is always sweet, and the Tommies were inclined to force an entrance. During a change of guard a Tommy who had his curiosity and initiative stimulated through recourse to arrick, the fiery liquor distilled from dates, stole into the most holy mosque in Kerbela. By a miracle he was got out unharmed, but for a few hours a general uprising with an attendant massacre of unbelievers was feared.

The great mosque lost much of its dignity through an atrocious clock-tower standing in the courtyard in front of it. It had evidently been found too expensive to cover this tower with a golden scale to shine in the sun, so some ingenious architect hit upon the plan of papering it with flattened kerosene-tins. It must have glinted gloriously at first, but weather and rain had rusted the cans and they presented but a sorry spectacle. From the thousand and one uses to which these oil-cans have been put by the native, one is inclined to think that the greatest benefit that has been conferred on the natives by modern civilization is from the hands of the Standard Oil Company.

There were a fair number of Indians living in Kerbela before the war, for devout Shiahs are anxious to be buried near the martyred sons of Ali, and when they are unable to move to Kerbela in their lifetime they frequently make provisions that their remains may be transported thither. The British found it a convenient abode for native rulers whom they were forced to depose but still continued to pension.

Hilleh, which stands near the ruins of ancient Babylon, is a modern town very much like Museyib. I never had a chance to study the ruins at any length. Several times we went over the part that had been excavated by the Germans immediately before the war. I understand that this is believed to be the great palace where Belshazzar saw the handwriting on the wall.[29] It is built of bricks, each one of which is stamped in cuneiform characters. There are very fine bas-reliefs of animals, both mythical and real. In the centre is the great stone lion, massively impressive, standing over the prostrate form of a man. The lion has suffered from fire and man; there have even been chips made in it recently by Arab rifles, probably not wantonly, but in some skirmish. Standing alone in its majesty in the midst of ruin and desolation amid the black tents of a people totally unable to construct or even appreciate anything of a like nature, it gave one much to think over and moralize about. The ruins of Babylon have been excavated only in very small part; there are great isolated mounds which have never been touched, and you can still pick up in the sand bits of statuary, and the cylinders that were used as seal-rings. The great city of Seleucia on the Tigris was built largely with bricks and masonry brought by barge from the ruins of Babylon through the canal that joined the two rivers.

The prophecy of Isaiah (13:19-22) has fallen true:

And Babylon, the glory of kingdoms, the beauty of the Chaldees' excellency, shall be as when God overthrew Sodom and Gomorrah.

It shall never be inhabited, neither shall it be dwelt in from generation to generation: neither shall the Arabian pitch tent there; neither shall the shepherds make their fold there.

But wild beasts of the desert shall lie there; and their houses shall be full of doleful creatures; and owls shall dwell there, and satyrs shall dance there.

And the wild beasts of the islands shall cry in their desolate houses, and dragons in their pleasant palaces: and her time is near to come, and her days shall not be prolonged.

---

29   Book of Daniel, Chapter 5. Daniel sees and interprets the hand writing on the wall. Mene, Mene, Tekel, Upharsin.

The Garden of Eden

Picture of the Ishtar gate.

Lion of Babylon

A few days after Christmas we were ordered to return to Baghdad. The going was still bad. We had a Ford tender in advance to find and warn us of the softest spots. Once it got into the middle of such a bottomless bog that, after trying everything else, I hit upon the idea of rolling it out. It was built all enclosed like a bread-van, and we turned it over and over until we had it clear of the mud. We had hard work with the heavy cars—sometimes we could tow one out with another, but frequently that only resulted in getting the two stuck. Once when the cars were badly bogged I went to a near-by Arab village to get help. I told the head man that I wanted bundles of brush to throw in front of the cars in order to make some sort of a foundation to pass them over. He at once started turning out his people to aid us, but after he had got a number of loads under way he caught sight of one of his wives, who, instead of coming to our assistance, was washing some clothes in a copper caldron by the fire. There followed a scene which demonstrated that even an Arab is by no means always lord of his own household. The wife refused to budge; the Arab railed and stormed, but she went calmly on with her washing, paying no more attention to his fury than if he were a fractious, unreasonable child. At length, driven to a white heat of rage, the head man upset the caldron into the fire with his foot. The woman, without a word, got up and stalked into a near-by hut, from which she refused to emerge. There was nothing for her discomfited adversary to do but go on with his rounds.

By manoeuvring and digging and towing we managed to make seven miles after fourteen hours' work that first day. Night found us close beside an Arab village, from which I got a great bowl of buffalo milk to put into the men's coffee. Early in the morning we were off again. The going was so much better that we were able to make Baghdad at ten o'clock in the evening.

The War In

Lt. General Sir Stanley Maude, KCB CMG, DSO, portrait, and below, his memorial and grave.

CHAPTER FOUR

SKIRMISHES AND RECONNAISSANCES ALONG THE KURDISH FRONT

We spent a few days making repairs and outfitting before starting off again. This time our destination was Deli Abbas, the headquarters of the Thirteenth Division. The town is situated in the plains below the foothills of the Persian Mountains, on the banks of the Khalis Canal, some seventy miles northeast of Baghdad. At dawn we passed out of the north gate, close to where General Maude is buried, and whirled across the desert for thirty miles to Bakuba, a prosperous city on the banks of the Diyala. From the junction of the greater Zab down to Kurna, where the Euphrates joins, this stream is the most important affluent of the Tigris. It was one of those bright, sparkling mornings on which merely to be alive and breathe is a joy. We passed a number of caravans, bringing carpets and rugs from Persia, or fruit and vegetables from the rich agricultural district around Bakuba. The silks manufactured here are of a fine quality and well known throughout the country.

After passing the big aerodrome near the town, the going became very bad; we struggled along through the village of Deltawa, in and out of unfathomable ditches. The rivers were in flood, and we ran into lakes and swamps that we cautiously skirted. Dark overtook us in the middle of a network of bogs, but we came upon an outpost of Welsh Fusiliers and spent the night with them. We had smashed the bottom plate of one of the cars, so that all the oil ran out of the crank-case, but with a side of the ever-useful kerosene tin we patched the car up temporarily and pushed off at early dawn. Our route wound through groves of palms surrounding the tumble-down tomb of some holy man, occasional collections of squalid little huts, and in the intervening "despoblado" we would catch sight of a jackal crouching in the hollow or slinking off through the scrub. Deli Abbas proved a half-deserted straggling town which gave evidence of having once seen prosperous days. Some Turkish aeroplanes heralded our arrival.

In front of us rose the Jebel Hamrin—Red Hills—beyond them the snow-clad peaks of the Kurdish Range. A few months previous we had captured the passes over the Jebel, and we were now busy repairing and improving the roads—in particular that across the Abu Hajjar, not for nothing named by the Arabs the "Father of Stones." Whenever the going permitted we went out on reconnaissances—rekkos, as we called them. They varied but slightly; the one I went on the day after reaching Deli Abbas might serve as model. We started at daybreak and ran to a little village called Ain Lailah, the Spring of Night, a lovely name for the small clump of palm trees tucked away unexpectedly in a hollow among barren foot-hills. There we picked up a surveyor—an officer whose business it was to make maps for the army. We passed through great herds of camels, some with small children perched on their backs, who joggled about like sailors on a storm-tossed ship, as the camels made away from the cars. There were villages of the shapeless black tents of the nomads huddled in among the desolate dunes. We picked up a Turk deserter who was trying to reach our lines. He said that his six comrades had been killed by Arabs. Shortly afterward we ran into a cavalry patrol, but the men escaped over some very broken ground before we could satisfactorily come to terms with them. It was lucky for the deserter that we found him before they did, for his shrift would have been short. We got back to camp at half past eight, having covered ninety-two miles in our windings—a good day's work.

Each section had two motorcycles attached to it—jackals, as one of the generals called them, in apt reference to the way in which jackals accompany a lion when hunting. The cyclists rode ahead to spy out the country and the best course to follow. When we got into action they would drop behind, and we used them to send messages back to camp. The best motor-cyclist we had was a Swiss named Milson. He was of part English descent, and came at once from Switzerland at the outbreak of the war to enlist. When he joined he spoke only broken English but was an exceedingly intelligent man and had been attending a technical college. I have never seen a more skilful rider; he could get his cycle along through the mud when we were forced to carry the others, and no one was more cool and unconcerned under fire. The personnel of the battery left nothing to be desired. One was proud to serve among such a fine set of men. Corporal Summers drove the car in which I usually rode, and I have never met with a better driver or one who understood his car so thoroughly, and possessed that intangible sympathy with it which is the gift of a few, but can be never attained.

We were still in the rainy season. We had to travel as light as possible, and all we could bring were forty-pounder tents, which correspond to the American dog-tent. Very low, they withstood in remarkable fashion the periodical hurricanes of wind and rain. They kept us fairly dry, too, for we were careful to ditch them well. There was room for two men to sleep in the turret of a Rolls, and they could spread a tarpaulin over the top to keep the rain from coming in through the various openings. The balance of the men had a communal tent or slept in the tenders. The larger tents in the near-by camps blew down frequently, but with us it happened only occasionally. There are happier moments than those spent in the inky blackness amid a torrential deluge, when you try to extricate yourself from the wet, clinging folds of falling canvas.

Time hung heavily when the weather was bad, and we were cooped up inside our tents without even a hostile aeroplane to shoot at. One day when the going was too poor to take out the heavy cars, I set off in a tender to visit another section of the battery that was stationed thirty or forty miles away in the direction of Persia, close by a town called Kizil Robat. We had a rough trip, with several difficult fords to cross. It was only through working with the icy water above our waists that we won through the worst, amid the shouts of "Shabash, Sahib!" ("Well done!") from the onlooking Indian troops. I reached the camp to find the section absent on a reconnaissance, for the country was better drained than that over which we were working. A few minutes later one of the cyclists came in with the news that the cars were under heavy fire about twenty-five miles away and one of them was badly bogged. I immediately loaded all the surplus men and eight Punjabis from a near-by regiment into the tenders. We reached the scene just after the disabled car had been abandoned. Some of the Turks were concealed in a village two hundred and fifty yards away; the rest were behind some high irrigation embankments. The free car had been unable to circle around or flank them because of the nature of the terrain. The men had not known that the village was occupied and had bogged down almost at the same time that the Turks opened fire. By breaking down an irrigation ditch the enemy succeeded in further flooding the locality where the automobile was trapped. The Turks made it hot for the men when they tried to dig out the car. The bullets spattered about them. It was difficult to tell how many Turks we accounted for. As dark came on, the occupants of the disabled car abandoned it and joined the other one, which was standing off the enemy but had lost all four tires and was running on its rims. We held a consultation and decided to stay where we were until dawn. We had scarcely made the decision when one of our cyclists arrived with orders from the brigade commander to return immediately. Although exceedingly loath to leave the armored car, we had no other course than to obey.

It was after midnight by the time we made back to camp. We were told that a small attack had been planned for the morning, and that then we could go out with the troops and recover our car, using some artillery horses to drag it free. The troops soon began filing past, but we didn't pull out till three o'clock, by which time we were reinforced by an armored car from another battery. We were held back behind the advanced cavalry until daylight, and felt certain that the Turks would have either destroyed or succeeded in removing our car. Nor were we wrong, for just as we breasted the hill that brought the scene of yesterday's engagement into view, we saw the smoke of an explosion and the men running back into the village. We cleared the village with the help of a squadron of the Twenty-First cavalry, and found that the car had been almost freed during the night. It was a bad wreck, but we were able to tow it. I wished to have a reckoning with the village head man, and walked to an isolated group of houses a few hundred yards to the left of the village. As I neared them a lively fusillade opened and I had to take refuge in a convenient irrigation ditch. The country was so broken that it was impossible for us to operate, so we towed the car back to camp.

Our section from Deli Abbas was moved up to take the place of the one that had been engaged, which now returned to Baghdad. We were camped at Mirjana, a few miles north of Kizil Robat, on the Diyala River. A pontoon bridge was thrown across and the cars were taken over to the right bank, where we bivouacked with a machine-gun company and a battalion of native infantry. The bed of the river was very wide, and although throughout the greater part of the year the water flowed only through the narrow main channel, in the time of the spring floods the whole distance was a riotous yellow torrent. We had no sooner got the cars across than the river began to rise. During the first night part of the bridge was carried away, and the rest was withdrawn. The rise continued; trees and brush were swept racing past. We made several fruitless attempts to get across in the clumsy pontoons, but finally gave it up, resigning ourselves to being marooned. We put ourselves on short rations and waited for the river to fall. If the Turks had used any intelligence they could have gathered us in with the greatest ease, in spite of our excellent line of trenches. On the fourth day of our isolation the river subsided as rapidly as it had risen.

We had good patrolling conditions, and each day we made long circuits. Sometimes we would run into a body of enemy cavalry and have a skirmish with them. Again we would come upon an infantry outpost and manoeuvre about in an effort to damage it. The enemy set traps for us, digging big holes in the road and covering them over with matting on which they scattered dirt to make the surface appear normal. The nearest town occupied by the Turks was Kara Tepe, distant from Mirjana eight or ten miles as the crow flies. In the debatable land were a number of native villages, and such inhabitants as remained in them led an unpleasantly eventful existence. In the morning they would be visited by a Turkish patrol, which would be displaced by us in our rounds. Perhaps in the evening a band of wild mountainy Kurds would blow in and run off some of their few remaining sheep. Then the Turks would return and accuse them of having given us information, and carry off some hostages or possibly beat a couple of them for having received us, although goodness knows they had little enough choice in the matter. There was one old sheik with whom I used often to sit and gossip while an attendant was roasting the berries for our coffee over the nearby fire. He was ever asking why we couldn't make an advance and put his village safely behind our fines, so that the children could grow fat and the herds graze unharmed. In this country Kurdish and Turkish were spoken as frequently as Arabic, and many of the names of places were Turkish—such as Kara Tepe, which means Black Mountain, and Kizil Robat, the Tomb of the Maidens. My spelling of these names differs from that found on many maps. It would be a great convenience if some common method could be agreed upon. At present the map-makers conform only in a unanimous desire to each use a different trans-literation

Kizil Robat is an attractive town. I spent some pleasant mornings wandering about it with the mayor, Jameel Bey, a fine-looking Kurdish chieftain of the Jaf tribe. He owned a lovely garden with date-palms, oranges, pomegranates, and figs. Tattered Kurds were working on the irrigation ditches, and a heap of rags lying below the wall in the sun changed itself into a small boy, just as I was about to step on it. Jameel's son was as white, with as rosy cheeks, as any American baby.

Harry Bowen, brother-in-law of General Cobbe, was the political officer in charge of Kizil Robat. He spoke excellent Arabic and was much respected by the natives. His house was an oasis in which I could always look forward to a pleasant talk, an excellent native dinner, and some interesting book to carry off. Although the town was small, there were three good Turkish baths. One of them belonged to Jameel Bey, but, judging from the children tending babies while squatting in the entrance portico, was generally given over to the distaff side and its friends. The one which we patronized, while not so grand a building, had an old Persian who understood the art of massage thoroughly, and there was nothing more restful after a number of days' hard work with the cars.

In the end of February there passed through Kizil Robat the last contingent of our former Russian Allies. They were Cossacks—a fine-looking lot as they rode along perched on their small chunky saddles atop of their unkempt but hardy ponies. When Russia went out of the war they asked permission to keep on fighting with us. They were a good deal of a problem, for they had no idea whatever of discipline, and it was most difficult to keep them in hand and stop them from pillaging the natives indiscriminately. They had been completely cut off from Russia for a long time but were now on their way back. A very intelligent woman doctor and a number of nurses who had been with them were sick with smallpox in one of our hospitals in Baghdad. When they recovered they were sent to India, for it was not feasible to repatriate them by way of Persia. When the Russians first established connection with us, some armored cars were sent to bring in the Cossack general, whose name we were told was Leslie. We were unprepared to find that he spoke no English! It turned out that his ancestors had gone over from Scotland to the court of Peter the Great.[30]

Hauling out a badly bogged fighting car

A Mesopotamian garage

---

30   Alexander Leslie (1580-1661) was a Scottish soldier who went abroad, and served both Sweden and Russia.

CHAPTER FIVE
THE ADVANCE ON THE EUPHRATES

Early in March we got orders to return to Baghdad, where all the armored cars were to be concentrated preparatory to an attack on the Euphrates front. There was much speculation as to our mission. Some said that we were to break through and establish connection with General Allenby's forces in Palestine. While I know nothing about it authoritatively, it is certain that if the state of affairs in France had not called for the withdrawal from the East of all the troops that could be spared, the attack that was launched in October would have taken place in March. We could then have advanced up the Euphrates, and it would have been entirely practical to cross over the desert in the cars by way of Tadmor.

When we got word to come in, the roads were in fearful shape and the rain was falling in torrents, but we were so afraid that we might miss the attack that we salvaged everything not essential and started to fight our way through the mud. It was a slow and wearisome process, but we managed to get as far as Bakuba by evening. The river was rising in one of its periodical floods and we found that the pontoon bridge had been cut half an hour before our arrival. No one could predict how long the flood would last, but the river rarely went down sufficiently to allow the bridge to be replaced within a week. At that time the railroad went only as far as Bakuba, and crossed the river on a wooden trestle, so I decided to try to load the motors on a flat car and get across the Diyala in that way.

After having made arrangements to do this I wandered off into the bazaar to get something to eat. In native fashion I first bought a big flap of bread from an old woman, and then went to a pickle booth to get some beets, which I wrapped in my bread. Next I proceeded to a meat-shop and ordered some lamb kababs roasted. The meat is cut in pellets, spitted on rods six or eight inches long, and lain over the glowing charcoal embers. In the shop there are long tables with benches beside them. The customer spreads his former purchases, and when his kababs are ready he eats his dinner. He next proceeds to a coffee-house, where he has a couple of glasses of tea and three or four diminutive cups of coffee to top off, and the meal is finished. The Arab eats sparingly as a rule, but when he gives or attends a banquet he stuffs himself to his utmost capacity.

Next morning we loaded our cars successfully and started off by rail for Baghdad, some thirty miles away. The railroad wound across the desert, with here and there a water-tank with a company from a native regiment guarding it. As we stopped at one particularly desolate spot, a young officer came running up and asked if we would have tea with him. He took us to his tent, where everything was ready, for he apparently always met the two trains that passed through daily. Poor fellow, he was only a little over twenty, and desperately lonely and homesick. Many of the young officers who were wounded in France were sent to India with the idea that they could be training men and getting on to the methods of the Indian army while yet recuperating and unfit to go back to the front. They were shipped out with a new draft when they had fully recovered. This boy had only been a month in the country, and ten days before had been sent off in charge of his Sikh company to do this wearisome guard duty.

We spent a few days in Baghdad refitting. The cars were to go out camouflaged to resemble supply-trucks, for every precaution was taken to prevent the Turks from realizing that we were massing men for an attack. The night before we were to start, word came in that the political officer at Nejef had been murdered, and the town was in revolt. We were ordered to send a section there immediately, so Lieutenant Ballingal's was chosen, while the rest of us left next morning with the balance of the battery for Hit. The first part of the route lay across the desert to Falujah, a prosperous agricultural town on the Euphrates. Rail-head lies just beyond at a place known as Tel El Dhubban—the "Hill of the Flies." From there on supplies were brought forward by motor transport, or in Arab barges, called shakturs. We crossed the river on a bridge of boats and continued up along the bank to Ramadie. Here I stayed over, detailed to escort the army commander on a tour of inspection.

The smaller towns along the Euphrates are far more attractive than those on the Tigris.

The country seems more developed, and most inviting gardens surround the villages. Hit, which lies twenty miles up-stream of Ramadie, is an exception. It is of ancient origin and built upon a hill, with a lovely view of the river. It has not a vestige of green on it, but stands out bleak and harsh in contrast to the palm groves fringing the bank. The bitumen wells near by have been worked for five thousand years and are responsible for the town being a centre of boat manufacture. With the bitumen, the gufas and mahelas are "pitched without and within," in the identical manner in which we are told that the ark was built. The jars in which the women of the town draw water from the river, instead of being of copper or earthenware as elsewhere, are here made of pitched wicker-work. The smell of the boiling bitumen and the sulphur springs is trying to a stranger, although the natives regard it as salubrious, and maintain that through it the town is saved from cholera epidemics. We had captured Hit a few weeks previously, and the aeroplanes flying low over the town had reported the disagreeable smell, attributing it to dirt and filth. "Eyewitness," the official newspaper correspondent, mentioned this in despatches, and when I was passing through, a proclamation of apology was being prepared to soothe the outraged and slandered townsfolk.

After taking the army commander back to rail-head, we retraced our steps with all speed to Hit, and thence the eight miles up-stream to Salahiyeh. The road beyond Hit was in fearful shape, and the engineers were working night and day to keep it open and in some way passable. In the proposed attack we were to jump off from Salahiyeh, and it was here that the armored cars were assembled. Our camp was close to a Turkish hospital. There were two great crescents and stars laid out for a signal to warn our aeroplanes not to drop bombs. One of the crescents was made of turf and the other of limestone. The batteries took turns in making the reconnaissances, in the course of which they would come in for a good deal of shelling. The road was unpleasant, because the camels and transport animals that had been killed during the Turkish retreat from Hit were by now very high. For some unknown reason there were no jackals or vultures to form a sanitary section. After reconnoitring the enemy positions and noting the progress they were making in constructing their defenses, we would make a long circuit back to camp.

A water-wheel on the Euphrates

One unoccupied morning I went over to an island on the river. Its cool, restful look had attracted me on the day I arrived, and it quite fulfilled its promise. Indeed, it was the only place I came across in Mesopotamia that might have been a surviving fragment of the Garden of Eden. It was nearly a mile long, and scattered about on it were seven or eight thick-walled and well-fortified houses. The entire island was one great palm-grove, with pomegranates, apricots, figs, orange-trees, and grape-vines growing beneath the palms. The grass at the foot of the trees was dotted with blue and pink flowers. Here and there were fields of spring wheat. The water-ditches which irrigated the island were filled by giant water-wheels, thirty to fifty feet in diameter. These "naurs" have been well described in the Bible, and I doubt if they have since been modified in a single item. There are sometimes as many as sixteen in a row. As they scoop the water up in the gourd-shaped earthenware jars bound to their rims, they shriek and groan on their giant wooden axles.

On the night of March 25 we got word that the long-expected attack would take place next morning. We had the cars ready to move out by three. Since midnight shadowy files had been passing on their way forward to get into position. One of our batteries went with the infantry to advance against the main fortified position at Khan Baghdadi. The rest of us went with the cavalry around the flank to cut the Turks off if they tried to retreat up-stream. We were well on our way at daybreak. The country was so broken up with ravines and dry river-beds that we knew we had a long, hard march ahead of us. Our maps were poor. A German officer that we captured had in some manner got hold of our latest map, and noting that we had omitted entirely a very large ravine, became convinced that any enveloping movement we attempted would prove a failure. As it happened, we came close to making the blunder he had anticipated, for we started to advance down to the river along the bank of a nullah which would have taken us to Khan Baghdadi instead of eight or ten miles above it, as we wished. I think it was our aeroplanes that set us straight. I was in charge of the tenders with supplies and spares, and spent most of the time in the leading Napier lorry. Occasionally I slipped into an armored car to go off somewhere on a separate mission. The Turks had doubtless anticipated a flanking movement and kept shelling us to a certain extent, but we could hear that they were occupying themselves chiefly with the straight attacking force. By afternoon we had turned in toward the river and our cavalry was soon engaged. The country was too broken for the cars to get in any really effective work. By nightfall we hoped we were approximately where we should be, and after making our dispositions as well as the circumstances would permit, we lay down beside the cars and were soon sound asleep. At midnight we were awakened by the bullets chipping the rocks and stones among which we were sleeping. A night attack was evidently under way, and it is always an eerie sensation. We correctly surmised that the Turks were in retreat from Khan Baghdadi and had run into our outposts. In a few minutes we were replying in volume, and the rat-tat-tats of the machine-guns on either side were continuous. The enemy must have greatly overestimated our numbers, for in a short time small groups started surrendering, and before things had quieted we had twelve hundred prisoners. The cavalry formed a rough prison-camp and we turned in again to wait for daylight.

At dawn we started to reconnoitre our position to find out just how matters stood. We came upon a body of two thousand of the enemy which had been held up by us in the night and had retreated a short distance to wait till it became light before surrendering. Among them were a number of German officers. They were all of them well equipped with machine-guns and rifles. Their intrenching tools and medical supplies were of Austrian manufacture, as were also the rolling kitchens. These last were of an exceedingly practical design. While we were taking stock of our capture we got word that Khan Baghdadi had been occupied and a good number of prisoners taken. We were instructed to press on and take Haditha, thirty miles above Khan Baghdadi. It was hoped that we might recapture Colonel Tennant, who was in command of the Royal Flying Corps forces in Mesopotamia. He had been shot down at Khan Baghdadi the day before the attack. We learned from prisoners that he had been sent up-stream immediately, on his way to Aleppo, but it was thought that he might have been held over at Haditha or at Ana.

We found that a lot of the enemy had got by between us and the river and had then swung back into the road. We met with little opposition, save from occasional bands of stragglers who concealed themselves behind rocks and sniped at us. Numbers surrendered without resistance as we caught up with them. We disarmed them and ordered them to walk back until they fell in with our cavalry, or the infantry, which was being brought forward in trucks. As we bowled along in pursuit the scene reminded me of descriptions in the novels of Sienkiewicz or Erckmann-Chatrian. The road was littered with equipment of every sort, disabled pack-animals, and dead or dying Turks. It was hard to see the wounded withering in the increasing heat—the dead were better off. We reached the heights overlooking Haditha to find that the garrison was in full retreat. Most of it had left the night before. Those remaining opened fire upon us, but in a halfhearted way, that was not calculated to inflict much loss. Many of the inhabitants of the town lived in burrows in the hillsides. Some of these caves had been filled with ammunition. The enemy had fired all their dumps, and rocks were flying about. We endeavored to save as much of the material as was possible. We were particularly anxious to get all papers dealing with the Arabs, to enable us to check up which were our friends and which of the ones behind our lines were dealing treacherously with us. We recaptured a lot of medical equipment and some ammunition that had been taken from our forces during the Gallipoli campaign.

Haditha is thirty-five miles from Khan Baghdadi, and Ana is an equal distance beyond. It was decided that we should push on to a big bridge shown on the map as eight miles this side of Ana. We were to endeavor to secure this before the Turks could destroy it, and cross over to bivouac on the far side. The road was in fair shape. Many of the small bridges were of recent construction. We soon found that our map was exceedingly inaccurate. Our aeroplanes were doing a lot of damage to the fleeing Turks, and as we began to catch up with larger groups we had some sharp engagements. The desert Arabs hovered like vultures in the distance waiting for nightfall to cover them in their looting.

That night we camped near the bridge. At dusk the Red Cross ambulances and some cavalry caught up. The latter had had a long, hard two days, with little to eat for the men and less for the horses, but both were standing up wonderfully. They were the Seventh Hussars and just as they reached us we recaptured one of their sergeants who had been made prisoner on the previous night. He had covered forty miles on foot, but the Turks had treated him decently and he had come through in good shape. We always felt that the Turk was a clean fighter. Our officers he treated well as long as he had anything to give or share with them. With the enlisted men he was not so considerate, but I am inclined to think that it was because he was not accustomed to bother his head much about his own rank and file, so it never occurred to him to consider ours. The Turkish private would thrive on what was starvation issue to our men. The attitude of many of the Turkish officers was amusing, if exasperating. They seemed to take it for granted that they would be treated with every consideration due an honored guest. They would complain bitterly about not being supplied with coffee, although at the time we might be totally without it ourselves and far from any source of supply. The German prisoners were apt to cringe at first, but as soon as they found they were not to be oppressed became arrogant and overbearing. At different times we retook men that had been captives for varying lengths of time. I remember a Tommy, from the Manchesters, if I am not mistaken, who had been taken before Kut fell, but had soon after made his escape and lived among the Kurdish tribesmen for seven or eight months before he found his way back to us. Quite a number of Indians who had been set to work on the construction of the Berlin-to-Baghdad Railway between Nisibin and Mosul made good their escape and struggled through to our lines.

It was a great relief when the Red Cross lorries came in and we could turn over the wounded to them. All night long they journeyed back and forth transporting such as could stand the trip to the main evacuation camp at Haditha.

By daybreak we were once more under way. Under cover of darkness the Arabs had pillaged the abandoned supplies, in some cases killing the wounded Turks. The transport animals of the enemy and their cavalry horses were in very bad shape. They had evidently been hard put to it to bring through sufficient fodder during the wet winter months when the roads were so deep in mud as to be all but impassable. Instead of being distant from Ana the eight miles that we had measured on the map, we found that we were seventeen, but we made it without any serious hindrance. The town was most attractive, embowered in gardens which skirt the river's edge for a distance of four or five miles. In addition to the usual palms and fruit-trees there were great gnarled olives, the first I had seen in Mesopotamia, as were also the almond-trees. It must be of great antiquity, for the prophet Isaiah speaks of it as a place where kings had reigned, but from which, even in his time, the grandeur had departed.

The greater part of the enemy had already abandoned the town, but we captured the Turkish governor and a good number of the garrison, and many that had escaped from Haditha. The disaster at Khan Baghdadi had only been reported the afternoon before, as we had of course cut all the telegraph wires, and the governor had not thought it possible we would continue the pursuit so far. He had spent most of his life in Hungary and had been given this post only a few months previous to our advance. From the prisoners we had taken at Haditha we had extracted conflicting estimates as to the time when Colonel Tennant, the commander of our air forces, had been sent on, and from those we took at Ana we received equally varying accounts. The cars had been ordered to push on in search of the colonel as long as sufficient gasolene remained to bring them back. Captain Todd with the Eighth Battery was in the lead when some thirty miles north of Ana they caught sight of a group of camels surrounded by horsemen. A couple of belts from the machine-guns scattered the escort, and Colonel Tennant and his companion, Major Hobart, were soon safe in the turret of one of the cars.

From some of our Turkish captives we heard about a large gold convoy which had been sent back from Ana; some said one day, and others two, before our arrival. The supply of fuel that we had brought in the tenders was almost exhausted, so that it would be necessary to procure more in order to continue the pursuit. Major Thompson, who was in command of the armored-car detachment, instructed me to take all the tenders and go back as far as was necessary to find a petrol dump from which I could draw a thousand gallons. I emptied the trucks and loaded them with such of the wounded as could stand the jolting they were bound to receive because of the speed at which I must travel. I also took a few of the more important prisoners, among them the governor of Ana. He was a cultivated middle-aged man who spoke no Arabic but quite good French. It was mid-afternoon when we started, and I hadn't the most remote idea where I would find a sufficient quantity of petrol. During the run back we were sniped at occasionally by Turks who were still hiding in the hills. A small but determined force could have completely halted the cars in a number of different places where the road wound through narrow rock-crowned gorges, or along ledges cut in the hillside and hemmed in by the river. In such spots the advance of the armored cars could cither have been completely checked, or at all events seriously hampered and delayed, merely by rolling great boulders down on top of us.

When we had retraced our steps for about sixty miles I was lucky enough to get wind of an enemy petrol dump that our men had discovered. It was a special aeroplane supply and the colonel of the infantry regiment who was guarding it had been instructed to allow none of it to be used for automobiles. He showed his desire to co-operate and his ability to read the spirit rather than the letter of a command by letting me load my tenders. The L.A.M. batteries were well regarded and we everywhere encountered a willingness to meet us more than half-way and aid us in the thousand and one points that make so much difference in obtaining results.

By the time that we had everything in readiness for our return run it was long after dark and the men were exhausted. I managed to get some tea, but naturally no sugar or milk. The strong steaming brew served to wash down the scanty supply of cold bully beef. Fortunately it was a brilliant starlit night, but even so it was difficult to avoid ditches and washouts, and the road seemed interminable. Not long after we left we ran into a couple of armored cars that had been detailed to bring the rescued aviators back, after they had been reoutfitted and supplied as far as our limited resources would permit. During the halt I found that my sergeant had produced from somewhere or other an emergency rum ration which he was issuing. An old-army, experienced sergeant always managed to hold over a reserve from former issues for just such occasions as tliis, when it would be of inestimable value. I had been driving all day and had the greatest difficulty in keeping awake. Twice I dozed off. Once I awakened just as the car started over the edge of an embankment; the other time a large rock in the road brought me back to the world. It was two o'clock in the morning when we wearily crept into Ana.

The expedition to capture the gold convoy was to start at four, so after two hours' sleep I bundled into one of the Rolls-Royces and the column swung out into the road. Through the mist loomed the sinister, businesslike outlines of the armored car ahead of me. Captain Carr of the Thirteenth L. A. M. B.'s[31] was in command of the expedition. Unless we were in action or in a locality where we momentarily expected to be under fire from rifle or machine-gun, the officer commanding the car and his N. C. O. stood in the well behind the turret, steadying themselves with leather loops riveted to its sides. On long runs the tool-boxes on either side of the well formed convenient seats. When the car became engaged the crew would get inside, pulling the steel doors shut. The slits through which the driver and the man next him looked could be made still smaller when the firing was heavy, and the peep-holes at either side and in the rear had slides which could be closed. The largest aperture was that around the tube of the gun. Splinters of lead came in continuously, and sometimes chance directed a bullet to an opening.

Captives on the shore, near Ana.

---

31   Light Armored Motor Battery.

One of our drivers was shot straight through the head near Ramadie. The bottom of the car was of wood, and bullets would ricochet up through it, but to have had it made of steel would have added too much weight. The large gasoline tank behind was usually protected by plating, but even so was fairly vulnerable. A reserve-tank holding ten gallons was built inside the turret. We almost invariably had trouble with the feed-pipes leading from it. During the great heat of the summer the inside of the turret was a veritable fiery furnace, with the pedals so hot that they scorched the feet.

Forty miles above Ana we came upon a large khan. These road-houses arc built at intervals along the main caravan routes. Their plan is simple: four walls with two tiers of rooms or booths built into them, enclosing an open court in which the camels and horses are tethered during the night. The whole is strongly made to resist the inroads of the desert tribesmen. As we drove to the heavy gate, a wild clamor met our ears from a confused jumble of Jewish and Armenian merchants that had taken refuge within. Some of them had left Ana on their way to Aleppo before the news of the fall of Khan Baghdadi had reached the town. Others had been despatched by the Turks when the news of our advance arrived. All had been to a greater or lesser degree plundered by the Arabs. Most of the baggage animals had been run off, and the merchants were powerless to move. The women were weeping and imploring help, and the children tumbled about among the confused heaps of merchandise. Some of the Armenians had relations in Baghdad about whom I was able to give them bits of information. All begged permission to go back to Ana and thence to the capital. We, of course, had no means of supplying them with transportation, and any attempt to recapture their lost property was out of the question.

A few miles on we made out a troop of Arabs hurrying inland, a mile or so away from us, across a couple of ravines. They had some of the stolen camels and were laden down with plunder. Two of our cars made a fruitless attempt to come to terms with them, but only succeeded in placing a few well-aimed bursts from their machine-guns among them.

We now began to come up with bands of Turks. We ran across a number of isolated stragglers who had been stripped by the Arabs. A few had been killed. They as a rule surrendered without any hesitation. We disarmed them and told them to walk back toward Ana. Several times we had short engagements with Turkish cavalry. As a general thing the ground was so very broken up that it was impossible to manoeuvre. I was riding a good deal of the time in the Ford tender that we had brought along with a few supplies, and when one of the tires blew out I waited behind to replace it. The armored cars had quite a start and we raced along to catch them. In my hurry I failed to notice that they had left the road in pursuit of a troop of cavalry, so when we sighted a large square building of the sort the Turks use as barracks, I made sure that the cars had been there before me. We drove up to the door and I jumped out and shoved it open. In the yard were some infantry and a few cavalry. I had only my stick—my Webley revolver was still in its holster. There was nothing to do but put on a bold front, so I shouted in Arabic to the man I took to be the officer in command, telling him to surrender, and trying to act as if our forces were just outside. I think he must have been more surprised than I was, for he did so immediately, turning over the post to me. Eldridge, the Ford driver, had succeeded in disengaging the rifle that he had strapped in beside him, and we made the rounds under the escort of our captive.

One wing of the post was used as a hospital, under the charge of an intelligent little Armenian. He seemed well informed about the war, and asked the question that was the universal wail of all the Armenians we encountered: "When would Great Britain free their country, and would she make it an independent state?" There was a definite limit to the number of prisoners we could manage to carry back, but I offered the doctor to include him. His answer was to go to his trunk and produce a picture of his wife and little daughter. They were, he told me, in Constantinople, and it was now two years since he had had leave, so that as his turn was due, he would wait on the chance of seeing his family.

When the cars came up we set off again in pursuit of the elusive gold convoy. We could get no accurate information concerning it. Some said it was behind, others ahead. We never ran it down. It may well be that it was concealed in a ravine near the road a few yards from where we passed. Just short of a town called Abu Kemal we caught three Germans. They were in terror when we took them, and afterward said that they had expected to be shot. Under decent treatment they soon became so insolent that they had to be brought up short.

During the run back to Ana we picked up the more important of our prisoners and took them with us. Twenty-two were all we could manage. I was running one of the big cars. It was always a surprise to see how easy they were to handle in spite of the weight of the armor-plate. We each took great pride in the car in which we generally rode. All had names. In the Fourteenth one section had "Silver Dart" and "Silver Ghost" and another "Gray Terror" and "Gray Knight." The car in which I rode a great deal of the time met its fate only a few days before the armistice, long after I had gone to France. Two direct hits from an Austrian "eighty-eight" ended its career.

It was after midnight when we got back to our camp in a palm-garden in Ana. Although we had not succeeded in capturing the gold convoy, we had brought in a number of valuable prisoners, and among other things I had found some papers belonging to a German political agent whom we had captured. These contained much information about the Arab situation, and through them it was all but proved that the German was the direct instigator of the murder of the political officer at Nejef. An amusing sidelight was thrown in the letters addressed by Arab sheiks through this agent to the Kaiser thanking him for the iron crosses they had been awarded. There must have been an underlying grim humor in distributing crosses to the Mohammedan Arabs in recognition of their efforts to withstand the advance into the Holy Land of the Christian invaders.

On our arrival at Ana we were told that orders had come through that the town be evacuated on the following morning. Preparations were made to blow up the ammunition dump, which was fortunately concentrated in a series of buildings that joined each other. We warned the inhabitants and advised them to hide in the caves along the hillsides. We ourselves went back to the camp which we had occupied near the bridge the night before entering Ana. During the afternoon Major Edye, a political officer, turned up, travelling alone with an Arab attendant. He pitched his camp, consisting of a saddle and blanket, close beside us. He was an extraordinarily interesting man, with a great gift for languages. In the course of a year or so's wandering in Abyssinia he had learned both ancient and modern Abyssinian. There was a famous German Orientalist with whom he corresponded in the pre-war days. He had mailed him a letter just at the outbreak, which, written in ancient Abyssinian, must have been a good deal of a puzzle to the censors.

The main explosion, taking place at the appointed time, was succeeded by smaller ones, which continued at gradually lengthening intervals throughout the night. General Cassels, who had commanded the cavalry brigade so ably throughout the advance, wished to return to Ana on the following morning in order to check up the thoroughness with which the dump had been destroyed. He took an escort of armored cars, and as I was the only one in the batteries who could speak Arabic, my services were requisitioned. As we approached the town the rattle of the small-arms ammunition sounded like a Fourth of July celebration. The general noticed that I had a kodak and asked me to go out into the dump and take some photographs. There was nothing to do but put on a bold front, but I have spent happier moments than those in which I edged my way gingerly over the smoking heaps to a ruined wall from which I could get a good view for my camera. As I came back a large shell exploded and we hastily moved the cars farther away.

I went to the mayor's house to find out how the town had fared. He was a solemn old Arab, and showed me the damage done by the shells with an absolutely expressionless face. The houses within a fair radius had been riddled, but the natives had taken our warning and no one had been killed. After a cup of coffee in a lovely garden on the river-bank.

I came back to the cars and we ran on through to Haditha. Here we were to remain for a week or ten days to permit the evacuation of the captured supplies.

Thus far we had been having good luck with the weather, but it now began to threaten rain. We crawled beneath the cars with our blankets and took such precautions as were possible, but it availed us little when a veritable hurricane blew up at midnight. I was washed out from under my car, but before dark I had marked down a deserted hut, and thither I groped my way. Although it was abandoned by the Arabs, living traces of their occupancy remained. Still, even that was preferable to the rain, and the roof proved unexpectedly water-tight.

All next day the storm continued. The Wadi Hauran, a large ravine reaching back into the desert for a hundred and fifty miles, became a boiling torrent. When we crossed over, it was as dry as a bone. A heavy lorry on which an anti-aircraft gun was mounted had been swirled away and smashed to bits. The ration question had been difficult all along, but now any further supply was temporarily out of the question.

Oddly enough, I was the only member of the brigade occupying Haditha who could speak enough Arabic to be of any use, so I was sent to look up the local mayor to see whether there was any food to be purchased. The town is built on a long island equidistant from either bank. We ferried across in barges. The native method was simpler. They inflated goatskins, removed their clothes, which they had fastened in a bundle on top of their heads, and with one hand on the goatskin they paddled and drifted over. By starting from the head of the island they could reach the shore opposite the down-stream end. The bobbing heads of the dignified old gray-beards of the community looked most ludicrous. On landing they would solemnly don their clothes, deflate the skins, and go their way.

The mayor proved both intelligent and agreeable. The food situation was such that it was obviously impossible for him to offer us any serious help. We held a conclave in the guest-house, sitting cross-legged among the cushions. In the centre a servant roasted coffee-beans on the large shovel-spoon that they use for that purpose. The representative village worthies impressed me greatly. The desert Arabs are always held to be vastly superior to their kinsmen of the town, and it is undoubtedly true as a general rule; nevertheless, the elders of Haditha were an unusually fine group of men. We got a few eggs, which were a most desirable luxury after a steady diet of black unsweetened tea and canned beef. We happened to have a sufficient supply of tea to permit us to make an appreciated gift to the village.

My shoes had collapsed a few days before and I borrowed a pair from a Turk who had no further use for them. These were several sizes too large and fashioned in an oblong shape of mathematical exactness. Even in the motor machine-gun service, there is little that exceeds one's shoes in importance, and I was looking forward with almost equal eagerness to a square meal and a pair of my own shoes. The supply of reading-matter had fallen very low. I had only Disraeli's *Tancred*, about which I found myself unable to share Lady Burton's feelings, and a French account of a voyage from Baghdad to Aleppo in 1808.[32] The author, Louis Jacques Rousseau, a cousin of the great Jean Jacques, belonged to a family of noted Orientalists.[33] Born in Persia, and married to the daughter of the Dutch consul-general to that country, he was admirably equipped for the distinguished diplomatic career that lay before him in the East and in northern Africa. His treatises on the archaeological remains that he met with on his many voyages are intelligent and thorough. The river towns have changed but little in the last hundred years, and the sketch of Hit might have been made only yesterday.

Within three days after the rise, the waters of the Wadi Hauran subsided sufficiently for us to cross, and I received orders to return to Baghdad. The rain had brought about a change in the desert since we passed through on our way up. The lines of Paterson, the Australian poet, kept running through my head:

32   Benjamin Disraeli (1804-1881) was a Tory MP, and served as Prime Minister from 1874-1880. He is the only Jewish prime minister. Lady Burton wrote of the book: 'Disraeli's *Tancred* and similar occult books were my favourites; but *Tancred*, with its glamour of the East, was the chief of them.' It's about a sensitive young man who goes to the Holy Land and falls in love with a jewish woman, gets kidnapped, and has a vision on Mount Sinai.

33   Jean-Baptiste-Louis-Jacques-Joseph Rousseau (1780-1831) served as consul in Basra, in Aleppo, and in Tripoli, and wrote several books on Persia, Iraq, Islam.

"For the rain and drought and sunshine make no changes in the street,
In the sullen line of buildings and the ceaseless tramp of feet,
But the bush hath moods and changes, as the seasons rise and fall,
And the men who know the bushland they are loyal through it all."[34]

The formerly arid floor of the desert was carpeted with a soft green, with myriads of little flowers, all small, but delicately fashioned.

There were poppies, dwarf daisies, expanses of buttercups, forget-me-nots, and diminutive red flowers whose name I did not know. It started raining again, and we only just succeeded in winning our way through to Baghdad before the road became impassable.

A "Red Crescent" ambulance

---

34   A.B. "Banjo" Paterson, CBE, (1864-1941) was a lawyer, journalist, novelist, and poet. He wrote "Waltzing Matilda" the unofficial anthem of Australia. Kermit Roosevelt mentions his other works, and meets him, later in this book, when he had been commissioned as an officer to lead the 2nd Remount Unit of the Australian Imperial Force. The poem is called 'In Defence of the Bush, and is printed in the front of the book.

CHAPTER SIX
BAGHDAD SKETCHES

Although never in Baghdad for long at a time, I generally had occasion to spend four or five days there every other month. The life in any city is complex and interesting, but here it was especially so. We were among a totally foreign people, but the ever-felt intangible barrier of color was not present. For many of the opportunities to mingle with the natives I was indebted to Oscar Heizer, the American consul. Mr. Heizer has been twenty-five years in the Levant, the greater part of which time he has spent in the neighborhood of Constantinople. The outbreak of the war found him stationed at one of the principal ports of the Black Sea. There he witnessed part of the terrible Armenian massacres, when vast herds of the wretched people were driven inland to perish of starvation by the roadsides. Quiet and unassuming, but ever ready to act with speed and decision, he was a universal favorite with native and foreigner alike.

With him I used to ferry across the river for tea with the Asadulla Khan, the Persian consul. The house consisted of three wings built around a garden. The fourth side was the river-bank. The court was a jungle of flowering fruit-trees, alive with birds of different kinds, all singing garrulously without pause. There we would sit sipping sherbet, and cracking nuts, among which salted watermelon seeds figured prominently. Coffee and sweets of many and devious kinds were served, with arrack and Scotch whiskey for those who had no religious scruples. The Koran's injunction against strong drink was not very conscientiously observed by the majority, and even those who did not drink in public, rarely abstained in private. Only the very conservative —and these were more often to be found in the smaller towns—rigorously obeyed the prophet's commands. It was pleasant to smoke in the shade and watch the varied river-craft slipping by. The public bellams plied to and fro, rowed by the swart owners, while against them jostled the gufas—built like the coracles of ancient Britain—a round basket coated with pitch. No Anglo-Saxon can see them without thinking of the nursery rhyme of the "wise men of Gotham who went to sea in a tub." These gufas were some of them twenty-five feet in diameter, and carried surprising loads—some-times sheep and cattle alone—sometimes men and women—often both indiscriminately mingled. Propelling a gufa was an art in itself, for in the hands of the uninitiated it merely spun around without advancing a foot in the desired direction. The natives used long round- bladed paddles, and made good time across the river. Crossing over in one was a democratic affair, especially when the women were returning from market with knots of struggling chickens slung over their shoulders.

Asadulla Khan's profile always reminded me of an Inca idol that I once got in Peru. Among his scribes were several men of culture who discoursed most sagely on Persian literature; on Sadi and Hafiz, both of whom they held to be superior to Omar Khayyam. I tried through many channels to secure a manuscript of the "Rubaiyat,"[35] but all I succeeded in obtaining was a lithograph copy with no place or date of publication; merely the remark that it had been printed during the cold months. I was told that the writings of Omar Khayyam were regarded as immoral and for that reason were not to be found in religious households. My Persian friends would quote at length from Sadi's *Gulistan* or *Rose Garden*, and go into raptures over its beauty.[36]

Below the consulate was a landing-place, and when we were ready to leave we would go down to the river-bank preceded by our servants carrying lanterns. They would call "Abu bellam" until a boat appeared. The term "abu" always amused me. Its literal meaning is "father." In the bazaars a shop-owner was always hailed as "father" of whatever wares he had for sale. I remember one fat old

---

35   Rubáiyát of Omar Khayyám was translated by Edward FitzGerald (1809-2883) and became widely popular in the United States and Great Britain. It's generally acknowledge that FitzGerald wrote most of the poems and passed them off as the writing of Omar Khayyam (1048-1131)

36   Saadi Shirazi (1210-1292) was a Persian poet. Voltaire enjoyed his works and nicknamed himself 'Saadi.' The poems reflected Sufi mysticism and romantic love. Iran has a national holiday dedicated to him.

man who sold porous earthenware jars—customers invariably addressed him as "Abu hub"— "Father of water-coolers."

My best friend among the natives was a Kurdish chief named Hamdi Bey Baban. His father had been captured and taken to Constantinople. After living there a number of years in semicaptivity he died—by poison it was said. Hamdi was not allowed to return to Kurdistan until after he was a grown man and had almost forgotten his native language. He spoke and read both French and English. Eventually permission was granted him to live in Baghdad as long as he kept out of the Kurdish hills, so he set off by motor accompanied only by a French chauffeur. Gasolene was sent ahead by camel caravan to be left for him at selected points. The journey was not without incident, for the villagers had never before seen an automobile and regarded it as a devil; often stones were thrown at them, and on one occasion they were mobbed and Hamdi only escaped by driving full speed through the crowd.

His existence in Baghdad had been subject to sudden upheavals. Once he was arrested and convoyed back to Constantinople; and just before the advance of the British his life was in great danger. Naturally enough he had little love for the Turk and staked everything on the final victory of the Allies.

He intended writing a book on the history of his family, in which he was much interested. For material he was constantly purchasing books and manuscripts. In the East many well-known histories still exist only in manuscript form, and when a man wishes to build up a library he engages scribes and sends them to the place where a famous manuscript is kept with an order to make a copy. In the same way Hamdi Bey had men busied transcribing rare chronicles dealing with the career of his family—extant in but one or two examples in mosques. lie once presented me with a large manuscript in Persian in which his family is mentioned, the mention taking the form of a statement to the effect that seventeen of them had had their heads removed!

Next to various small tradesmen with whom I used to gossip, drink coffee, and play dominoes, my best Arab friend was Abdul Kader Pasha, a striking old man who had been a faithful ally to the

British through thick and thin. The dinners at his house on the riverbank were feasts such as one reads of in ancient history. Course succeeded course without any definite plan; any one of them would have made a large and delicious meal in itself. True to Arab custom, the son of the house never sat down at table with his father, although before and after dinner he talked and smoked with us.

I had a number of good friends among the Armenians. There was not one of them but had some near relation, frequently a parent or a brother or sister, still among the Turks. Sometimes they knew them to be dead, more frequently they could only hope that such was the case and there was no further suffering to be endured. Many of these Armenians belonged to prominent families, numbering among their members men who had held the most important government posts in Constantinople. The secretary of the treasury was almost invariably an Armenian, for the race outstrips the Jews in its money touch.

With one family I dined quite often—the usual interminable Oriental feast varying only from the Arab or Turkish dinners in a few special national dishes. All, excepting the aged grandmother, spoke French, and the daughters had a thorough grounding in the literature. Such English books as they knew they had read in French translations. The house was attractively furnished, with really beautiful rugs and old silverware. The younger generation played bridge, and the girls were always well dressed in European fashion. Whence the clothes came was a mystery, for nothing could have been brought in since the war, and even in ante-bellum days foreign clothes of that grade could never have been stocked but must have been imported in individual orders. The evenings were thoroughly enjoyable, for everything was in such marked contrast to our every-day life. It must be remembered that these few Armenians were the only women with whom we could talk and laugh in Occidental fashion. By far the best-informed and cleverest Arab was Pere Anastase. He was a Catholic, and under the supervision of the Political Department edited the local Arab paper. All his life he had worked building up a library—gathering rare books throughout Syria and Mesopotamia. He was himself an author of no small reputation. Just before the British took Baghdad the Germans pillaged his collection, sending the more valuable books to Constantinople and Berlin, and turning the rest over to the populace. The soldiers made great bon-fires of many—others found their way to the bazaars, where he was later able to repurchase some of them. When talking of the sacking of his house, Pere Anastase would work himself into a white heat of fury and his eyes would flash as he bitterly cursed the vandals who had destroyed his treasures.

It was in Baghdad that I first ran into Major E. B. Soane, whose *Through Mesopotamia and Kurdistan in Disguise* is a classic.[37] Soane was born in southern France, his mother French and his fa-ther English. The latter walked across the United States from ocean to ocean in the early forties, so Soane came by his roving, adventur-ous spirit naturally. When still but little more than a boy he went out to work in the Anglo-Persian Bank, and immediately interested himself in the language and literature of the country. Some of his holidays he spent in the British Museum translating and catalogu-ing Persian manuscripts. Becoming interested in the Kurds, he spent a number of years among them, learning their languages and customs and joining in their raids.

As soon as we got a foothold in the Kurdish Hills, Soane was sent up to administer the captured territory. His headquarters were at Khanikin, twenty-five miles from Kizil Robat and but a short distance from the Persian frontier. One morning during the time that I was stationed in that district I motored over to see him. It was a glorious day. The cloud effects were most beautiful, towering in billows of white above the snow peaks, against a background of deepest blue. The road wound in and out among the barren foot-hills until suddenly as I topped a rise I saw right below a great clump of palm-trees, with houses showing through here and there—the whole divided by a lovely river bestridden by an old seven-arch bridge. I picked my way through the narrow streets, scattering ragged Kurds right and left; past part of the covered bazaar, until I came to a house with a large courtyard, thronged with a motley array of Kurdish irregulars, armed with every sort of

---

37   Ely Bannister Soane (1881-1923) was a linguist and intelligence agent. He worked in Persia, converted to Islam, and was arrested by the Turks at the beginning of the war.

weapon. It was there that Soane administered his stern but practical justice, for he thoroughly understood how to handle these men.

The district had suffered fearfully, for it had been occupied in turn by Turk and Russian, and then Turk again, before we took it over, and the unfortunate natives had been pillaged and robbed mercilessly.[38] Thousands starved to death. When I was at Deli Abbas ghastly bands of ragged skeletons would come through to us begging food and work. Soane turned a large khan on the outskirts of the town into a poorhouse, and here he lodged the starving women and children that drifted in from all over Kurdistan. It was a fearful assemblage of scarecrows. As they got better he selected women from among them to whom he turned over the administration of the khan. They divided the unfortunates in gangs, and supervised the issue of dates on which they were fed. Such as were physically able were employed in cleaning the town. The Kurds are a fine, self-respecting race and it was easy to understand Soane's enthusiasm for them.

In Baghdad you lived either in the cellars or on the housetops. The former were called serdabs. A large chimney, cowled to face the prevailing wind, served for ventilation, and on the hottest days one was cool and comfortable. We slept on the roofs, and often dined there, too. Since the town was the General Headquarters of the Expeditionary Force, one was always sure to meet many friends. A comfortable and well-run officers' club was installed, as well as warrant officers' and enlisted men's clubs.

Occasionally race meetings were planned and the various divisions would send representatives. Frank Wooton, the well-known jockey, was a despatch-rider, and usually succeeded in getting leave enough to allow him to ride some general's horses. An Arab race formed part of the programme. Once a wild tribesman who had secured a handsome lead almost lost the race by taking off his cloak and waving it round his head as he gave ear-piercing shouts of triumph. The Arab riding second was less emotional and attended better to the business in hand, but his horse was not quite good

---

38   Russia and Turkey fought over ten wars in this region since the 1500s. When World War One began, Russia invaded and had the upper hand until 1917, when the revolution caused the army to disintegrate.

enough to make the difference.

The scene at the race-course was a gay one. The color was chiefly contributed by the Jewesses who wore their hooded silk cloaks of lively hue—green or pink or yellow. The only crowd that I saw to vie with it was one which watched the prisoners taken at Ramadie march through the town. Turkish propaganda, circulated in the bazaars, gave out that instead of taking the prisoners we claimed, we had in reality suffered a defeat, and it was decided that the sight of the captive Turks would have a salutary effect upon the townsmen. Looking down from a housetop the red fezzes and the gay-colored abas made the crowd look like a vast field of poppies.

When I was at Samarra an amusing incident took place in connection with a number of officers' wives who were captured at Ramadie. The army commander didn't wish to ship them off to India and Burma with their husbands, so he sent them up to Samarra with instructions that they be returned across the lines to the Turks. After many aeroplane messages were exchanged it was agreed that we should leave them at a designated hill and that the Turks would later come for them. Meanwhile we had arranged quarters for them, trying to do everything in a manner that would be in harmony with the Turkish convenances. When the wives were escorted forth to be turned back to their countrymen, they were all weeping bitterly. Whether it was that the Turk in his casual manner decided that one day was as good as another, or whether he felt that he had no particular use for these particular women, we never knew, but at all events twenty-four hours later one of our patrols came upon the prisoners still forlornly waiting. We shipped them back to Baghdad.

Occasionally I would go to one of the Arab theatres. The plays were generally burlesques, for the Arab has a keen sense of humor and greatly appreciates a joke. Most of the puns were too involved for me to follow, but there was always a certain amount of slap-stick comedy that could be readily understood. Then there was dancing—as a whole monotonous and mediocre; but there was one old man who was a remarkable performer, and would have been appreciated on any stage in the world. The topical songs invariably amused me—they were so universal in spirit. The chorus of one which was a great hit ran: "Haido, haido, rahweni passak!" "I say, I say, show me your pass." There had been much trouble with spies and every one was required to provide himself with a certificate of good conduct and to show it on demand. It was to this that the song referred.

Captain C. G. Lloyd was my companion on many rambles among the natives. He had been stationed in Burma and India for many years, and was a good Persian scholar. Like every one who has knocked about to any extent among native peoples, his career had not been lacking in incident. I remember on one occasion asking him why it was that he never joined me in a cup of coffee when we stopped at a coffee-house. He replied that he had always been wary of coffee since a man with him was poisoned by a cup which was intended for him.

I always looked forward to a trip to Baghdad, for it gave me a chance to mingle in a totally different life from that which daily surrounded me, and temporarily, at least, forget about the war in which the world was plunged. Still, the morning set to leave invariably found me equally glad to shove off once more into the great expanses of the desert.

A jeweller's booth in the bazaar

CHAPTER SEVEN
THE ATTACK ON
THE PERSIAN FRONT

When I reached headquarters after the attack on the Euphrates front, I was expecting to hear that my transfer to France had gone through and receive orders to proceed thither immediately. It had always been my intention to try to join the American army once it began to take a real part in the war, and for some time past I had been casting about in my mind for the best method to carry out my plans. When affairs looked so very black for the Allied forces in March and April, 1918, I decided that France was the place where every one, who could by any possibility manage it, should be. General Gillman, the chief of staff, had on more than one occasion shown himself a good friend, and I determined to once more task his kindness. He said that he thought he could arrange for my transfer to France, and that once there I could work out the best way of getting into the American army.

Everything went well, and I was daily expecting my orders, when Major Thompson, who commanded the brigade of armored cars, sent for me and told me that an advance was being planned on the Kurdish front. Only two batteries were to be taken—the Eighth and the Thirteenth—but he said that he would like to have me go along in command of the supply-train. Of course I jumped at the chance, as the attack promised to be most interesting.

We were told to be ready to move on an hour's notice. For several days the weather held us back. The rain, helped out by the melting snow from the mountains, caused the rivers to rise in flood. The Tigris rose sixteen feet in a night. The lower bridge was broken and washed away. Everything possible was done to reinforce the upper bridge, but it was hourly expected to give way under the strain of the whirling yellow waters. The old Arab rivermen said that they could tell by the color just which of the tributaries were in spate. When they saw or thought they saw a new admixture, they would shake their heads and say: "Such and such a river is now also in flood—the Tigris will rise still further."

On the night of April 24 we at length got our orders and at six o'clock the following morning we set out, prepared to run through to Ain Leilah. The country was indeed changed since I passed through six weeks before. The desert had blossomed. We ran through miles and miles of clover; the sweet smell seemed so 'wholesomely American, recalling home and family, and the meadows of Long Island. The brilliant red poppies were more in keeping with the country; and we passed by Indian cavalry reinforcements with the scarlet flowers stuck in their black hair and twined in the head-stalls of the horses.

As we approached the hills they looked less bleak—a soft green clothed the hollows, and the little oasis of Ain Leilah no longer stood out in the same marked contrast as when last I visited it. The roads were in good shape, and we reached camp at four in the afternoon. I took one of the tenders and set off to look up some old friends in the regiments near by. As I passed a group of Arabs that had just finished work on the roads, I noticed that they were playing a game that was new to me. A stake was driven into the ground, with a horsehair rope ten or twelve feet in length attached to it. An old man had hold of the end of the rope. About the stake were piled some clothes, and the Arabs were standing around in a circle just out of reach of the man with the rope. The object was to dart in and snatch up something from the heap without the old man who was on guard catching you. They were enjoying themselves hugely—the oldest graybeards behaving as if they were children—a very pleasant side of the Arab.

Our instructions were to be ready to pull out before daybreak. The mission was, as usual, a flanking one. The direct attack was to be delivered on Kara Tepe, and, if that were successful, upon Kifri. We were to intercept the arrival of reinforcements, or cut off the retreat of the garrisons, as the case might be.

In the early morning hours the country was lovely—rolling

grass land "with a hint of hills behind"[39]—miles of daisies with clusters of blood-red poppies scattered through them— and occasional hollows carpeted with a brilliant blue flower. In the river courses there were numbers of brilliantly hued birds—the gayest colors I saw in Mesopotamia with the exception of the vivid arsenic-green birds around Ana on the Euphrates. In one place I thought that the ground was covered with red flowers, but a close inspection proved it to be myriads of tiny red insects swarming on the grass stems.

Column marching is slow and wearisome, and after the sun rose the heat became intense. The dust smothered us; there was not a breath of air to rid us of it for even a moment. The miles seemed interminable. At noon we halted beside a narrow stream known as Oil River—a common name in this part of the country where oil abounds and the water is heavily impregnated with it. For drinking it was abominable—and almost spoiled the tea upon which we relied for a staple. A few miles beyond, the engineers found a suitable location to throw a bridge across the creek. The main body was halted at a place known as Umr Maidan and we were sent over the bridge to form across the main road leading from Kara Tepe back into the Turkish territory.

It was nightfall before we had effected a crossing, and we groped our way along until we came upon the road. It was impossible to do very much in the way of selecting a position, but we arranged the cars as best we could. When you were oft at large in the desert you were what the army called "Out in the blue," and that was certainly our situation on the night of April 26. We all expected that we would intercept traffic going one way or the other, but the night passed without incident or excitement.

By four in the morning we were once more feeling our way along through the darkness. As it lightened we came under observation by the Turks, who started in to shell us. We learned from our aeroplanes that Kifri had been evacuated; the garrison was falling

39   Rudyard Kipling's poem, 'The Explorer.' "But at last the country altered - White Man's country past disrupting-/Rolling grass and open timber, with a hint of hills behind -/There I found me food and water, and I lay a week recruiting./Got my strength and lost my nightmares. Then I entered on my find."

back along a road running parallel to the one on which we were,
separated by eight or ten miles of broken country. By this time our
cavalry had caught up with us. They pushed off across country to
intercept the Turks. We attempted to do likewise but it was more
difficult, and what with dodging in and out to avoid a ravine here
or a hill there, we made little headway. At length we struck a road
that led in approximately the direction whither we wished to go. It
was already early afternoon before, upon topping a rise, we caught
sight of a good-sized body of Turks marching on a road which
ran along the base of a range of steep, stony hills. We put on as
much speed as was possible, and headed north to try to intercept
them. The cavalry were coming from the south, and while we were
circling around they charged in upon the Turks. It was a stirring
scene. The powerful Indians sat their horses with the utmost grace.
Their drawn sabres flashed in the sun. As they came to close quarters the turbaned heads bent forward and we could hear the shouts
and high-pitched cries of triumph as the riders slashed at the foe.
The wounded and dead testified to their skill as swordsmen. The
whole sight reminded me more of the battle books I read as a boy
than anything I saw in the war. About six hundred prisoners were
taken, but many of the Turks escaped to the mountains and lay
among the rocks, whence they could snipe at us with impunity.
They were a tenacious lot, for all next day when we were using the
road below the hills they continued to shoot at us from the places
whence it was impossible to dislodge them.

Indian cavalry bringing in prisoners after the charge

The War In

Above: British soldier feeding a surrendered Turkish soldier.
Below: Indian soldiers guarding Turkish captives.

While the prisoners were being brought in we caught sight of one of our aeroplanes crashing. Making our way over to it we found that neither the pilot nor the observer was seriously hurt. Flying in Mesopotamia was made unusually difficult by the climatic conditions. The planes were designed for work in France and during the summer months the heat and dryness warped the propeller blades and indeed all the wooden parts. Then, too, the fine dust would get into the machinery when the aviator was taxiing for a start. Many pilots coming out from France with brilliant records met an early and untimely end because they could not realize how very different the conditions were. I remember one poor young fellow who set off on a reconnaissance without the food and water he was required by regulations to carry. He got lost and ran out of gasolene—being forced to land out in the desert. The armored cars went off in search of him, and on the second morning after' he had come down they found his body near their bivouac. He had evidently got that far during the night and died of exhaustion and exposure practically within hearing. He was stripped of his clothes; whether this had been done by himself or by the tribesmen was never determined. A death of this sort always seems so much sadder than being legitimately killed in combat. The L.A.M. batteries were in close touch with the Royal Flying Corps, for when news came in that a plane was down in the desert or some part of the debatable land, we would be detailed to go out in search of the occupants. A notice printed in Arabic, Persian, Turkish, and Kurdish was fastened into each aeroplane informing the reader of the reward that would be paid him if the pilot were brought in safety to the British lines. This was done in case a plane got lost and was driven down out of its course among the tribesmen.

The night of the 27th we bivouacked once more "out in the blue."[40] Dawn found me on my way back to Umr Maidan to lay in a new supply of gasolene. I made a rapid trip and caught up with the armored cars in action in a large swampy plain. The grass was very high and the ground so soft that it was difficult to accomplish anything. Two or three small hills offered vantage-points, but they

40   Henry Van Dyke (1852-1933), from the poem 'Day and Night,' "For look, how the edge of the sky grows gray,/ While the stars die out in the blue above,/ And the wan moon fades away."

were not neglected by the Turk, and among those that fell was the colonel of the Twenty-First cavalry—the regiment that had acquitted itself so well in the charge of the day before.

We were ten miles from Tuz Kliurmartli, the next important town held by the enemy now that Kifri had been taken. It was thither that the Turks had been retreating when we cut them off. Finding that we were unable to operate effectively where we were, it was decided that we should make our way across to the Kifri-Kirkuk road and advance along it to make a frontal attack upon Tuz. Our orders were to proceed to a deserted village known as Kulawand, and wait there for the command to advance. When we got to the road we found the hills still occupied by camel-guns and machine-guns. We replied ineffectively, for we had no means of dislodging them, nor did the cavalry when they came up. Kulawand we found to be a fair-sized native village unoccupied save for a single hut full of old women and children. Here we waited until nightfall for the orders that never came. I sat under a ruined wall reading alternatively Camoens' *Lusiad* and *David Harum* until darkness fell.[41]

During the night some infantry came up, both native and British. They had had stiff marching during the last few days, and were done up, but very cheerful at the prospect of an attack on the morrow. They had some hard fighting ahead of them. The King's Own in particular distinguished itself in taking a stubbornly contested and strongly held hill.

At dawn we were under way. We had heard reports during the night that the Turks had evacuated Tuz—but it was not long before we found that such was not the case. They were still there and showed every evidence of staying. A small village five or six miles to the southwest was also bitterly contested. Our cavalry did some excellent work, capturing small hills held with machine-guns.

We advanced down the road beside the hills. A mile before reaching Tuz we ran into the Aq Su, a large stream flowing through a narrow cleft in the hills. Fortunately the river was very low, and

41   *David Harum* was an 1898 novel by retired banker Edward Noyes Westcott (1846-1898), about an honest country banker. It was a posthumous success and was adapted for the stage, radio, and screen.

there were several places where it was spread out over such a wide
bed that it seemed as if it might be possible to get the cars across.
I emptied a Ford van and set out to do some prospecting. First I
went up-stream, which was toward the mountains, but I could not
go far, for there was an ancient fort situated at the mouth of the
gorge, and it had not been evacuated. Finding a likely looking place
a little below, I made a cast and just succeeded in getting through.
It was easy to see that it would not be possible for the low-swung
Rolls to cross under their own power, for the fly-wheel would
throw the water up into the motor. There was nothing to do but
send back for artillery horses to pull the armored cars across.

Meanwhile, as our artillery had practically ceased firing on
the town and the Turks seemed to have entirely evacuated it, I
thought that I would go up and take over and see whether there
had not been some valuable documents left behind. I drove along
past some abandoned artillery into the main street. A number of
Turkish soldiers came up to surrender and I told them to have the
Reis Beledia—the town mayor—report to me. When he came I directed him to take me to the quarters of the Turkish commanding
general. As we drove through the covered bazaar everything was
closed. Scarcely anybody was in the streets—but I could see the
inhabitants peeping out from behind lattices. It was a good thing
to have the old mayor along, for he served as an excellent hostage,
and I kept close watch upon him. He brought me to a prosperous,
neat-looking house with heavy wooden doors. In response to his
summons an old woman came and ushered us into a large, cool
room, well furnished and with beautiful Kurdish rugs. There we
found four young girls, who, it was explained to me, formed the
Turkish general's "field harem." He had left in too much of a hurry
to take them with him. They were Kurds and Circassians, or Geor-
gians—and the general had shown no lack of taste in his selection!
True to the tradition of the Garden of Eden, this harem proved
disastrous to a brother officer who, having heard of my capture,
sent me "priority" over the field service lines a ribald message as to
its disposition. "Priority" wires are sent only on affairs of the greatest importance, and when I left the country my friend was slated
to explain matters before a court martial. There were no papers of

any great value to be found, and I told the mayor to take me to the more important ammunition and supply dumps. By the time I had located these some cavalry had come in, and I went back to the river to help get the fighting cars across.

Once we had these safely over we set out in pursuit of the Turks. The next town of importance was a ramshackle mud-walled affair called Tauq, twenty miles beyond, on the far side of a river known as the Tauq Chai. The leading cars pursued to within sight of the town and came in for a good deal of shelling.

The Turks we captured were in far poorer shape than those we had recently taken on the Euphrates front. Their shoes were worn out, they were very ragged, and, what was of greater significance, they were badly nourished. The length of their line of communications had evidently severely strained them. Supplies had to come overland all the way from Nisibin, which is more than a hundred miles beyond Mosul. The broken country made the transportation a difficult problem to solve. It was a miracle that they had the morale to fight as they did under such disadvantageous conditions.

Here, as throughout the campaign, it was a continual source of pride to see the way in which our soldiers behaved to the natives. I never heard of a case in which man, woman, or child was wrongfully treated. Minor offenses were sometimes committed, but these were quickly righted. No doubt there were isolated instances of wrong-doing, for in such a large army there are bound to be degenerate individuals from whose conduct it is unfair to judge the whole.

That night we encamped in the outskirts of Tuz, not far from the Turkish aerodrome. Next morning one of the batteries was ordered to reconnoitre as far as the town—pursuing a different route than that taken on the previous day. The commanding officer asked me to go along because of my knowledge of Arabic. The road followed the telegraph-lines, and part of the time that was the only way in which we could distinguish it from the surrounding country. Of course, the map was hopelessly incorrect. The villages were not even rightly named. A great deal of reconnoitring was called for, and in one village we had to knock the corner off a mud house to enable us to make a sharp right-angle turn. The natives were in pitiful condition. The Turks had not only taken all their crops, but even the grain that should be reserved to sow for the following year. The sheep had been killed in the lambing season, so the flocks were sadly depleted. Such standing grain as there was left looked flourishing. The wheat waved above the cars.

As we came out of a deep, broad ravine that had caused us much delay and difficulty, we caught sight of an attractive town situated on a steep, flat-topped hill. Upon drawing near, a fine-looking, white-bearded Arab rode up on a small gray mare. He said that he was the head man of the town; that he hated the Turks, and would like to be of any assistance possible to us. I asked him if the enemy had evacuated Tauq. He replied that they had. I then asked him if he were positive about it. He offered to accompany us to prove it. The trail was so bad that we could not go fast, and he rode along beside us at a hand-gallop.

The Kurd and his wife

Sheik Muttar and the two Kurds

When we came to the river in front of the town we found that it was impossible to get the armored cars across. The Turks had evidently fallen back, but not far, for they were dropping in shells with regularity. Our Arab friend told us that there was a bridge six miles up-stream, but it was too late for us to attempt it, and we turned back to Tuz after arranging with Sheikh Muttar to meet us in the morning.

Next day we found him waiting for us as he had promised. With him were two handsome Kurds. One of them had his wife perched behind him on the horse's crupper. Together they undertook to guide us up to the bridge. It was invariably difficult to find out from natives whether or not a road was passable for motor-cars. They were accustomed to think only in terms of horses or men, and could not realize that a bad washout might be impassable for automobiles. Curiously enough, even those natives whom we had taken along with us on several reconnaissances as guides could not be trusted to give an opinion as to the feasibility of a proposed route. We experienced no little trouble in following our guides to the bridge, although we afterward discovered a good road that cut off from the main trail about half-way between Tuz and Tauq.

When we reached the bridge we found it to be a solid, well-built affair of recent construction. The retreating Turks had tried to blow it up, but the most vital charges had failed to go off, so the damage done would not be sufficiently serious to stop our passage, after six or seven hours' preliminary work. We immediately sent back for the engineers, and put in the time while waiting by taking a much needed bath in the rapids beneath one of the side arches. Every one who has wandered about in the waste places of the world can re-call certain swims that will always stand out in his memory. Perhaps they have been after a long and arduous hunt—perhaps at the end of a weary march. Our plunge in the Tauq Chai took its place among these.

In the late afternoon we drove back to Tuz. Our camp there was anything but cheerful, for swarms of starving townsfolk hovered on the outskirts ready to pounce on any refuse that the men threw away. Discarded tin cans were cleaned out until the insides shone like mirrors. The men gave away everything they could possibly spare from their rations. As the news spread, the starving mountain Kurds began straggling in; and the gruesome band made one glad to leave camp early and return after dark. Our line of communication was so extended that it was impossible to attempt any relief work.

The following morning we crossed over the bridge with little trouble, but ran into a lot of difficulty when we tried to make our way down to the town. A couple of miles above the main town there is a small settlement grouped on a hill around the mosque of Zain El Abidin. The "mutabelli," or keeper of the shrine, is an important personage in the community, so when he appeared riding a richly caparisoned stallion and offered to accompany us to the town, we welcomed the opportunity of going in under such good auspices. We decided to take Seyid Mustapha, for that was his name, in one of the Ford vans with us. It was comparatively easy to get the light car up over the precipitous, rocky trail; and eventually one of the fighting cars succeeded in following. I was driving, with Mustapha beside me. In front of us on a white horse galloped the Seyid's attendant singing and shouting and proclaiming our arrival. We stopped at Mustapha's house for a cup of coffee and a discussion of events. The information which we secured from him afterward proved unusually correct. I took him on with us to the town so that he could identify the head man and see that we got hold of the right people. Our reception was by no means cordial, although after we had talked a little and explained what we were after, the mayor became cheerful and expansive. He had a jovial, rotund face, covered in large part by a bushy beard, and would have done excellently as a model for Silenus. In the town were a handful of Turkish stragglers—among them a stalwart Greek who spoke a little English. He said that he had been impressed into service by the Turks and was most anxious to join our forces.

We found large stores of ammunition and other supplies, among them a wireless set. What interested us most, I am afraid, was the quantity of chickens that we saw strutting about. A few of them and a good supply of eggs found their way to the automobiles in short order. We were always very particular about paying for whatever we took, and seeing that the men did likewise; our reputation went before us, and the native, as a rule, took it for granted that we would pay. It was up to the officers to see that the prices were not exorbitant. We always used Indian currency—the rupee and the anna. In normal times a rupee is about a third of a dollar. Throughout the occupied area Turkish currency also circulated, but the native invariably preferred to be paid in Indian. Curiously enough, even on entering towns like Tauq, we found the inhabitants eager for payment in rupees. I was told that in the money market in Bagh-dad a British advance would be heralded by a slump in Turkish exchange. Paper rupees were almost everywhere as readily accepted as silver, but paper liras and piasters were soon of so little value that they were no longer in circulation.

When we got back to camp I found a ware informing me that I had been transferred to the American army, and ordering me to report at once to Baghdad to be sent to France. Major Thompson asked me if I would delay my return until the end of the advance. It was rumored that we would continue to push on and would attack Kirkuk. Many felt that the difficulty that was already being experienced in rationing us would preclude our thrusting farther. Still, I made up my mind that as long as the major wished it and would wire for permission I would stay a few days longer on the chance of the attack continuing.

On the morning of the 3rd we moved camp to the far side of the Tauq Chai bridge. When the tenders were unloaded I started back to bring up a supply of gasolene, with the purpose of making a dump in case we were called upon for a further advance. I was told that the nearest supply from which I could draw was at Umr Maidan; and the prospect of running back, a distance of seventy miles, was not cheerful. When I got as far as Tuz I found a friend in charge of the dump there, and he let me draw what I wanted, so I turned back to try to get to the bridge by dark. One car after another got in trouble; first it was a puncture, then it was a tricky carburetor that refused to be put to rights; towing-ropes were called into requisition, but the best had been left behind, and those we had were rotted, and broke on every hill. Lastly a broken axle put one of the tenders definitely out of commission, and, of course,

I had to wait behind with it. To add to everything, a veritable hurricane set in, with thunder and lightning and torrents of rain. The wind blew so hard that I thought the car would be toppled over. What made us more gloomy than anything else was the thought of all the dry river courses that would be roaring floods by morning, and probably hold up the ration supply indefinitely.

Two days later the orders for which we had been waiting came through. We were to march upon a town called Taza Khurmatli, lying fifteen miles beyond Tauq and ten short of Kirkuk. If we met with no opposition there we were to push straight on. From all we could hear Taza was occupied only by cavalry, which would probably fall back without contesting our advance. The cars had been out on reconnaissance near the town for the last two days, and had come in for artillery and machine-gun fire; but it was believed that the Turks had everything ready to withdraw their guns on our approach.

In the gray light that preceded dawn we saw shadowy columns of infantry and artillery and cavalry passing by our camp. The costumes of the different regiments made a break in the drab monotony. The Mesopotamian Expeditionary Force was composed of varied components. Steel helmets could be worn only in winter. In many of the native regiments the British officers wore tasselled pugrees, and long tunics that were really shirts, and an adaption of the native custom of wearing the shirt-tails outside the trousers. The Gurkhas were supplied with pith helmets. It was generally claimed that this was unnecessary, but the authorities felt that coming from a cold, high climate they would be as much affected by the Mesopotamian sun as were Europeans. The presence of the Indian troops brought about unusual additions to the dry "General Routine Orders" issued by general headquarters. One of them, referring to a religious festival of the Sikhs, ran:

"The following cable message received from Sunder Signh Hagetha, Amritsar, addressed to Sikhs in Mesopotamian force:

"To our most Dear Brothers now serving the Benign King-Emperor oversea, the chief Khalsa Dewan tenders hearty and sincere greetings on the auspicious Gurpurb of First Guru. You are upholding

*the name and fame of Gurupurb. Our hearts are with you and our
prayers are that Satguru and Akalpurkh may ever be with you and
lead you to victory and return home safe, after vanquishing the
King-Emperor's foes, with honor and flying colors."*

The British Empire was well and loyally served by her Indian
subjects, and by none more faithfully than the Sikhs.

We let the column get well started before we shoved off in our
cars. The trail was wide enough to pass without interfering; and
long before we were in sight of Taza we had taken our place ahead.
As was foreseen, the enemy evacuated the town with scarce a show
of resistance. I set off to interview the local head man. In the spring
all the upper Mesopotamian towns are inundated by flocks of
storks, but I have never seen them in greater force than in Taza. On
almost every housetop were a couple, throwing their heads back
and clattering their beaks in the odd way that gives them their on-
omatopoetic Arabic name of Lak-Lak. It sounded like the rattle of
machine-guns; so much so that on entering the village, for the first
second I thought that the Turks were opening up on us. No native
will molest a stork; to do so is considered to the last degree inaus-
picious.

There was but little water in the river running by Taza, and we
managed to get the cars through under their own power. A few
miles farther on lay a broad watercourse, dry in the main, but with
the centre channel too deep to negotiate, so there was nothing to
be done without the help of the artillery horses. The Turks were
shelling the vicinity of the crossing, so we drew back a short distance and sent word that we were held un waiting for assistance to
get us over.

Once we had reached the far side we set out to pick our way
round Kirkuk to get astride the road leading thence to Altun Kupri.
This is the main route from Baghdad to Mosul, the chief city on
the upper Tigris, across the river from the ruins of Nineveh. It was
a difficult task finding a way practicable for the cars, as the ground
was still soft from the recent rains. It was impossible to keep defi-
laded from Turkish observation, but we did not supply them with
much in the way of a target. At length we got round to the road,

and started to advance down it to Kirkuk. The town, in common with so many others in that part of the country, is built on a hill. The Hamawand Kurds are inveterate raiders, and good fortifications are needed to withstand them. As we came out upon the road we caught sight of our cavalry preparing to attack. The Turks were putting up a stout resistance, with darkness fast coming to their aid. After approaching close to the town, we were ordered to return to a deserted village for the night, prepared to go through in the early morning.

The co-ordinates of the village were given, and we easily found it on the map; but it was quite another proposition to locate it physically. To add to our difficulties, the sky clouded over and pitchy blackness settled down. It soon started to rain, so we felt that the best we could do was select as likely a spot as came to hand and wait for morning. I made up my mind that the front seat of a van, uncomfortable and cramped as it was, would prove the best bed for the night. My estimate was correct, for at midnight the light drizzle, that was scarcely more than a Scotch mist, turned into a wild, torrential downpour that all but washed away my companions. The waterproof flap that I had rigged withstood the onslaughts of wind and rain in a fashion that was as gratifying as it was unexpected. The vivid flashes of lightning showed the little dry ravine beside us converted into a roaring, swirling torrent. The water was rushing past beneath the cars, half-way up to their hubs. A large field hospital had been set up close to the banks of the stream at Taza. We afterward heard that the river had risen so rapidly that many of the tents and a few ambulances were washed away.

By morning it had settled down into a steady, businesslike downpour. We found that we were inextricably caught in among some low hills. There was not the slightest chance of moving the fighting cars; they were bogged down to the axle. There was no alternative other than to wait until the rain stopped and the mud dried. Fortunately our emergency rations were still untouched.

Our infantry went over at dawn, and won through into the town. If it had not been for the rain we would have made some important captures. As it was, the Turks destroyed the bridge across the Hasa Su and retreated to Altun Kupri by the road on the farther bank. From a hill near by we watched everything, powerless to help in any way.

At noon the sky unexpectedly cleared and the sun came out. We unloaded a Ford van, and with much pushing and no little spade work managed to get it down to a road running in the direction of Kirkuk. We found the surface equal to the light car, and slowly made our way to the outskirts of the town, with occasional halts where digging and shoving were required. We satisfied ourselves that, given a little sun, we could bring the armored cars out of their bog and through to the town.

Kirkuk

Next morning, in spite of the fact that more rain had fallen during the night, I set to work on my tenders, and at length succeeded in putting them all in Kirkuk. We were billeted in the citadel, a finely built, substantial affair, with a courtyard that we could turn into a good garage. The Turks had left in great haste, and, although they had attempted a wholesale destruction of everything that they could not take, they had been only partially successful. In my room I found a quantity of pamphlets describing the American army—with diagrams of insignia, and pictures of fully equipped soldiers of the different branches of the service. There was also a map of the United States showing the population by States. The text was, of course, in Turkish and the printing excellently done. What the purpose might be I could not make out.

The wherefore of another booklet was more obvious. It was an illustrated account of alleged British atrocities. Most of the pictures purported to have been taken in the Sudan, and showed decapitated negroes. Some I am convinced were pictures of the Armenian massacres that the Turks had themselves taken and in a thrifty moment put to this useful purpose. This pamphlet was printed at the press in Kirkuk.

There were a number of excellent buildings —mainly workshops and armories, but the best was the hospital. The long corridors and deep windows of the wards looked very cool. An up-to-date impression was given by the individual patient charts, with the headings for the different diagnoses printed in Turkish and French. The doctors were mainly Armenians. The occupants were all suffering from malnutrition, and there was a great deal of starvation in the town.

I did not wish to return to Baghdad until I could be certain that we were not going to advance upon Altun Kupri. The engineers patched up the bridge, and we took the cars over to the other side and went off on a reconnaissance to ascertain how strongly the town was being held. The long bridge from which it gets its name could easily be destroyed, and crossing over the river would be no light matter. The surrounding mountains limited the avenue of attack. Altogether it would not be an easy nut to crack, and the Turks had evidently determined on a stand. What decided the army commander not to make any further attempt to advance was most probably the great length of the line of communications, and the recent floods had made worse conditions which were bad enough at the best. The ration supply had fallen very low, and it seemed impossible to hold even Kirkuk unless the rail-head could be advanced materially.

I put in all my odd moments wandering about the bazaars. The day after the fall the merchants opened their booths and transacted business as usual. The population was composed of many races, chiefly Turcoman, Kurd, and Arab. There were also Armenians, Chaldeans, Syrians, and Jews. The latter were exceedingly prosperous. Arabic and Kurdish and Turkish were all three spoken. Kirkuk is of very ancient origin—but of its early history little is known. The natives point out a mound which they claim to be Daniel's tomb. Two others are shown as belonging to Shadrach and Meshech; that of the third of the famous trio has been lost. There are many artificial hills in the neighborhood, and doubtless in course of time it will prove a fruitful hunting-ground for archaeologists. As far as I could learn no serious excavating has hitherto been undertaken in the vicinity.

The bazaars were well filled with goods of every sort. I picked up one or two excellent rugs for very little, and a few odds and ends, dating from Seleucid times, that had been unearthed by Arab laborers in their gardens or brick-kilns. There were some truck-gardens in the outskirts, and we traded fresh vegetables for some of our issue rations. There are few greater luxuries when one has been living on canned foods for a long time. I saw several ibex heads nailed up over the doors of houses. The owners told me that they were to be found in the near-by mountains, but were not plentiful. There is little large game left in Mesopotamia, and that mainly in the mountains. I once saw a striped hyena. It is a nocturnal animal, and they may be common, although I never came across but the one, which I caught sight of slinking among the ruins of Istabulat, south of Samarra, one evening when I was riding back to camp. Gazelle were fairly numerous, and we occasionally shot one for venison. It was on the plains between Kizil Robat and Kara Tepe that I saw the largest bands. Judging from ancient bas-reliefs lions must at one time have been very plentiful. In the forties of the last century Sir Henry Layard speaks of coming across them frequently in the hill country; and later still, in the early eighties, a fellow countryman, Mr. Fogg, in his *Land of the Arabian Nights*, mentions that the English captain of a river steamer had recently killed four lions, shooting from the deck of his boat.[42] Rousseau speaks of meeting, near Hit, a man who had been badly mauled by a lion, and was going to town to have his wounds cared for. Leopards

42   William Perry Fogg (1826-1909) was an American writer and traveler, and the inspiration for Phileas Fogg in Verne's novel *Around the World in 80 Days*.

and bears are to be met with in the higher mountain regions, and wild boars are common in many districts. They inhabit the thickets along the river-banks, in country that would permit of much sound sport in the shape of pig-sticking.

Game-birds are found in abundance; both greater and lesser bustard; black and gray partridges, quail, geese, duck, and snipe. A week's leave could be made provide good shooting and a welcome addition to the usual fare when the wanderer returned. Every sort of shotgun was requisitioned, from antiquated muzzle-loaders bought in the bazaar to the most modern creations of Purdy sent out from India by parcel-post.

After waiting a few days further, to be certain that an attack would not be unexpectedly ordered, I set out on my return trip to Baghdad. The river at Taza was still up, but I borrowed six mules from an accommodating galloping ambulance, and pulled the car across. We went by way of Kifri, a clean, stone-built town that we found all but empty. The food situation had become so critical that the inhabitants had drifted off, some to our lines, others to Persia, and still others to Kirkuk and Mosul. Near Kifri are some coal-mines about which we had heard much. It is the only place in the country where coal is worked, and we were hoping that we might put it to good use. Our experts, however, reported that it was of very poor quality and worth practically nothing.

CHAPTER EIGHT
BACK THROUGH PALESTINE

Several days later I embarked at Baghdad on one of the river boats. I took Yusuf with me to Busra to put me aboard the transport for Egypt. It was the first time he had ever been that far downstream, and he showed a fine contempt for everything he saw, comparing it in most disparaging terms to his own desolate native town of Samarra. The cheapness, variety, and plenty of the food in the bazaars of Busra were the only things that he allowed in any way to impress him.

I was fortunate enough to run into some old friends, and through one of them met General Sutton, who most kindly and opportunely rescued me from the dreary "Rest-Camp" and took me to his house. While I was waiting for a chance to get a place on a transport, he one morning asked me to go with him to Zobeir, where he was to dedicate a hospital. Zobeir is a desert town of ten thousand or so inhabitants, situated fifteen miles inland from Busra. The climate is supposed to be more healthful, and many of the rich and important residents of the river town have houses there to which they retire during the summer months. To an outsider any comparison would seem only a refinement of degrees of suffocation. The heat of all the coastal towns of the Persian Gulf is terrific.

Zobeir is a desert town, with its ideals and feelings true to the inheritance of the tribesmen. It is a market for the caravans of central Arabia. A good idea of the Turkish feeling toward it may be gathered from the fact that the inhabitants were exempt from military service. This was a clear admission on the part of the Turk that he could not cope with the situation, and thought it wisest not to attempt something which he had no hope of putting through. It was, therefore, a great triumph for the British and a sure wedge into the confidence of the desert folk when the hospital was opened, for any people that can introduce so marked an innovation among the hidebound desert communities must have won their confidence and respect in a remarkable degree. Ibrahim, the hereditary Sheikh of Zobeir, himself contributed largely to the fund for the endowment. It was arranged that Doctor Borrie, who among his other duties ran the civil hospital at Basra, should periodically include Zobeir in his rounds. The Sheikh showed us over the building. It was cool, comfortable, and very sanitary. The Indian who was to be resident physician had every appearance of intelligence and proficiency. Old Ibrahim gave us a large banquet of the orthodox type. There was a sheep roasted whole, and dishes of every sort of meat and vegetable marshalled upon the table, which fairly groaned beneath their weight. We had innumerable speeches. General Sutton made an excellent address, which an interpreter translated into Arabic. Our Arabian hosts were long-winded, and the recognized local orator was so classical in his phrases and forms and tenses that it was impossible to do more than get the general drift of what he said. Luckily I had in my pocket a copy of the *Lusiads*, which I surreptitiously read when the speeches became hopelessly long drawn out.

I was allotted space' on a British India/boat, the *Torrilla*, that was to take to Egypt afield artillery regiment of the Third Division. As we dropped down-stream and I watched a disconsolate Yusuf standing on the dock, I felt that another chapter had closed—an interesting one at that. I was not left long to muse on what the next would bring forth before there was a cry of "fire"; and from where I was standing in the smoking-room I could see, through the open hatchways, the soldiers hurrying about below decks. As the ship was well ballasted with ammunition, anything that happened would take place quickly, and only those on the spot could hope to control events, so I stayed where I was. A few minutes later the fire was reported out.

The long two weeks' trip through the Persian Gulf and round to the Red Sea was monotonously peaceful. Being "unattached," I had no regular duties. Occasionally I attended "stables," and wandered around the horse lines. The great heat below decks had far less effect upon the horses than would be supposed. Of course, they were well cared for, and many were seasoned veterans that had taken more than one long sea voyage. If I am not mistaken, only one was lost on the trip.

Most of the time I lay back in my rhoorkhee chair and read whatever I could find in the ship's library. The wireless broke down a few days after we left Busra, so we got no news whatever of the outer world, and soon ceased to speculate on what might be happening in France.

At length, on the morning of June 4, we dropped anchor in Suez harbor. We had hoped that the *Torrilla* would run through the canal to Port Said, but the disembarkation officer told us that we were all to be unloaded at Suez and proceed by rail. When I reached Alexandria I learned that a convoy had just sailed and there would not be another for two weeks at earliest. Sir Reginald Wingate, who had long been a family friend, was the British High Commissioner. Lady Wingate and he with the utmost hospitality insisted on my moving out to the residency to wait for my sailing.

When I left for Mesopotamia Lord Derby had given me a letter to General Allenby which I had never had an opportunity to present. Sir Reginald suggested that I could not do better than make use of this enforced delay by going up to Palestine. The railway was already running to Jerusalem and you could go straight through from Cairo with but one change. At Kantara you crossed the canal and entered the military zone. Leaving there at half past eleven in the evening the train reached Ludd, which was general headquarters, at seven the following morning.

Every one that I had ever met who knew General Allenby was wildly enthusiastic about him, and you had only to be with him a few minutes to realize how thoroughly justified their enthusiasm was. He represented the very highest type of the British soldier, and more need not be said. On the morning on which I arrived an attack was in progress and we could hear the drumming of the guns. The commander-in-chief placed a car at my disposal and I went around visiting old friends that I had made in Mesopotamia or still earlier in England, before the war. Among the latter was Colonel Ronald Storrs, the military governor of Jerusalem. With him I spent several days. Life in the Holy City seemed but little changed by the war. There was an interesting innovation in the Church of the Nativity at Bethlehem. The different Christian religious sects, in particular the Greek and Latin Catholics, were prone to come to blows in the church, and bloodshed and death had more than once been the result. To obviate this it had been the custom to have a regular relief of Turkish soldiers stationed in the church. Their place was now taken by British and French and Italians. Each nationality in rotation furnished the guard for a day. At the festival of the distribution of the Sacred Fire from the Church of the Holy Sepulchre in Jerusalem there were usually a number of accidents caused by the anxiety to reach the portal whence the fire was given out. The commander-in-chief particularly complimented Colonel Storrs upon the orderly way in which this ceremony was conducted under his regime. The population of Jerusalem is exceedingly mixed—and the percentage of fanatics is of course disproportionately large. There are many groups that have been gathered together and brought out to the Holy Land with distinctly unusual purposes. One such always had an empty seat at their table and confidently expected that Christ would some day appear to occupy it. The long-haired Russian and Polish Jews with their felt hats and shabby frock coats were to be met with everywhere. In the street where the Jews meet to lament the departed glory of Jerusalem an incongruous and ludicrous element was added by a few Jews, their bowed heads covered with ancient derby hats, wailing with undefeated zeal.

It is a mournful fact that the one really fine building in Jerusalem should be the Mosque of Omar—the famous "Dome of the Rock." This is built on the legendary site of the temple of Solomon, and the mosaics lining the inside of the dome are the most beautiful I have ever seen. The simplicity is what is really most felt, doubly so because the Christian holy places are garish and tawdry, with tin-foil and flowers and ornate carving. It is to be hoped that the Christians will some day unite and clean out all the dreary offerings and knickknacks that clutter the Church of the Holy Sepulchre. Moslems hold the Mosque of Omar second in sanctity only to the great mosque in the holy city of Mecca. It is curious, therefore, that they should not object to Christians entering it. Mohammedans enter barefoot, but we fastened large yellow slippers over our shoes, and that was regarded as filling all requirements. Storrs pointed out to me that it was quite unnecessary to remove our hats, for that is not a sign of respect with Moslems, and they keep on their red fezzes. The mosque was built by the Caliph Abd el Melek, about fifty years after Omar had captured Jerusalem in 636 A. D.[43] Many of the stones used in building it came from the temple of Jupiter. In the centre lies the famous rock, some sixty feet in diameter, and rising six or seven feet above the floor of the mosque. To Mohammedans it is more sacred than anything else in the world save the Black Stone at Mecca. Tradition says that it was here that Abraham and Melchizedek sacrificed to Jehovah, and Abraham brought Isaac as an offering.[44] Scientists find grounds for the belief that it was the altar of the temple in the traces of a channel for carrying off the blood of the victims. The Crusaders believed the mosque to be the original temple of Solomon, and, according to their own reports, rededicated it with the massacre of more than ten thousand Moslems who had fled thither for refuge. The wrought-iron screen that they placed around the rock still remains. The cavern below is the traditional place of worship of many of the great characters of the Old Testament, such as David and Solomon and Elijah. From it Mohammed made his night journey to heaven, borne on his steed El Burak.[45] In the floor of the cavern is an opening covered with a slab of stone, and said to go down to the centre of the world and be a medium for communicating with the souls of the departed.

The military governor has been at work to better the sanitary conditions in Jerusalem. Hitherto the only water used by the townsfolk had been the rain-water which they gathered in tanks. Some years ago it was proposed to bring water to the city in pipes,

43   Omar (582-644), the second Rashidun Caliph, successor of Abu Bakr, during the expansionary conquest of early Islam, when they conquered much of the Middle East, North Africa, and Persia.
44   In Genesis 14, Abraham gives ten percent of the war spoils to Melchizedek. In Genesis 22, he binds Isaac on the altar.
45   The famous Night Journey of Mohammed describes a flight to Heaven. They are mentioned in Surah 17 and Surah 53 of the Koran.

some of which were already laid before the inhabitants decided that such an innovation could not be tolerated. The British have put in a pipe-line, and oddly enough it runs to the same reservoir whence Pontius Pilate started to bring water by means of an aqueduct. They have also built some excellent roads through the surrounding hills. Here, as in Mesopotamia, one was struck by the permanent nature of the improvements that are being made. Even to people absorbed in their own jealousies and rivalries the advantages that they were deriving from their liberation from Turkish rule must have been exceedingly apparent.

A street in Jerusalem

The situation in Palestine differed in many ways from that in Mesopotamia, but in none more markedly than in the benefits derived from the propinquity of Egypt. Occasional leaves were granted to Cairo and Alexandria and they afforded the relaxation of a complete change of surroundings. I have never seen Cairo gayer. Shepherd's Hotel was open and crowded—and the dances as pleasant as any that could be given in London. The beaches at Ramleh, near Alexandria, were bright with crowds of bathers, and the change afforded the "men from up the line" must have proved of inestimable value in keeping the army contented. There were beaches especially reserved for non-commissioned officers and others for the privates—while in Cairo sightseeing tours were made to the pyramids and what the guide-books describe as "other points of interest."

When I left Mesopotamia I made up my mind that there was one man in Palestine whom I would use every effort to see if I were held over waiting for a sailing. This man was Major A. B. Paterson, known to every Australian as "Banjo" Paterson. His two most widely read books are *The Man from Snowy River* and *Rio Grande's Last Race*;[46] both had been for years companions of the entire family at home and sources for daily quotations, so I had always hoped to some day meet their author. I knew that he had fought in the South African War, and I heard that he was with the Australian forces in Palestine. As soon as I landed I asked every Australian officer that I met where Major Paterson was, for locating an individual member of an expeditionary force, no matter how well known he may be, is not always easy. Every one knew him. I remember well when I inquired at the Australian headquarters in Cairo how the man I asked turned to a comrade and said: "Say, where's 'Banjo' now? He's at Moascar, isn't he?" Whether they had ever met him personally or not he was "Banjo" to one and all.

On my return to Alexandria I stopped at Moascar, which was the main depot of the Australian Remount Service, and there I found him. He is a man of about sixty, with long mustaches and strong aquiline features— very like the type of American plainsman that Frederic Remington so well portrayed. He has lived everything that he has written. At different periods of his life he has dived for pearls in the islands, herded sheep, broken broncos, and

---

46   See footnote 32.

known every chance and change of Australian station life. The Australians told me that when he was at his prime he was regarded as the best rider in Australia. A recent feat about which I heard much mention was when he drove three hundred mules straight through Cairo without losing a single animal, conclusively proving his argument against those who had contested that such a thing could not be done. Although he has often been in England, Major Paterson has never come to the United States. He told me that among American writers he cared most for the works of Joel Chandler Harris and O. Henry—an odd combination![47]

While in Egypt I met a man about whom I had heard much, a man whose career was unsurpassed in interest and in the amount accomplished by the individual. Before the war Colonel Lawrence was engaged in archaeological research under Professor Hogarth of Oxford University.[48] Their most important work was in connection with the excavation of a buried city in Palestine. At the outbreak of hostilities Professor Hogarth joined the Naval Intelligence and rendered invaluable services to the Egyptian Expeditionary Forces. Lawrence had an excellent grounding in Arabic and decided to try to organize the desert tribes into bands that would raid the Turkish outposts and smash their lines of communication. He established a body-guard of reckless semi-outlaws, men that in the old days in our West would have been known as "bad men." They became devoted to him and he felt that he could count upon their remaining faithful should any of the tribes with which he was raiding meditate treachery. He dressed in Arab costume, but as a whole made no effort to conceal his nationality. His method consisted in leading a tribe off on a wild foray to break the railway, blow up bridges, and carry off the Turkish supplies. Swooping down from out the open desert like hawks, they would strike once and be off before the Turks could collect themselves. Lawrence explained that he

---

47   O. Henry (1862-1910) wrote many well known short stories, such as "Gift of the Magi." Joel Chandler Harris (1848-1908) was a Southerner who created Uncle Remus and Br'er Rabbit stories.
48   i.e. T.E. Lawrence of Arabia, CB, DSO, (1888-1935) the famous soldier who helped lead the Arab revolt against Ottoman rule. He studied under the noted archaeologist D.G. Hogarth, CMG, FRGS, FBA, (1862-1927) who excavated sites in Crete, Egypt, and served in Naval Intelligence during the war.

had to succeed, for if he failed to carry off any booty, his reputation among the tribesmen was dead—and no one would follow him thereafter. What he found hardest on these raids was killing the wounded—but the dread of falling into the hands of the Turks was so great that before starting it was necessary to make a compact to kill all that were too badly injured to be carried away on the camels. The Turks offered for Colonel Lawrence's capture a reward of ten thousand pounds if dead and twenty thousand pounds if alive. His added value in the latter condition was due to the benefit that the enemy expected to derive from his public execution. No one who has not tried it can realize what a long ride on a camel means, and although Lawrence was eager to take with him an Englishman who would know the best methods of blowing up bridges and buildings, he could never find any one who was able to stand the strain of a long journey on camel back.

Lawrence told me that he couldn't last much longer, things had broken altogether too well for him, and they could not continue to do so. Scarcely more than thirty years of age, with a clean-shaven, boyish face, short and slender in build, if one met him casually among a lot of other officers it would not have been easy to single him out as the great power among the Arabs that he on every occasion proved him-self to be. Lawrence always greatly admired the Arabs—appreciating their many-sidedness—their virility—their ferocity—their intellect and their sensitiveness. I remember well one of the stories which he told me. It was, I believe, when he was on a long raid in the course of which he went right into the out-skirts of Damascus—then miles behind the Turkish lines. They halted at a ruined palace in the desert. The Arabs led him through the various rooms, explaining that each was scented with a different perfume. Although Lawrence could smell nothing, they claimed that one room had the odor of ambergris—another of roses—and a third of jasmine;—at length they came to a large and particularly ruinous room. "This," they said, "has the finest scent of all—the smell of the wind and the sun." I last saw Colonel Lawrence in Paris, whither he had brought the son of the King of the Hedjaz to attend the Peace Conference.

When I got back to Alexandria I found that the sailing of the convoy had been still further delayed. Three vessels out of the last one to leave had been sunk, involving a considerable loss of life. The channel leading from the harbor out to sea is narrow and must be followed well beyond the entrance, so that the submarines had an excellent chance to lay in wait for outgoing boats. The greatest secrecy was observed with regard to the date of leaving and destination—and of course troops were embarked and held in the harbor for several days so as to avoid as far as possible any notice being given to the lurking enemy by spies on shore.

The transports were filled with units that were being hurried off to stem the German tide in France, so casual officers were placed on the accompanying destroyers and cruisers. I was allotted to a little Japanese destroyer, the *Umi*.[49] She was of only about six hundred and fifty tons burden, for this class of boat in the Japanese navy is far smaller than in ours. She was as neat as a pin, as were also the crew. The officers were most friendly and did everything possible to make things comfortable for a landsman in their limited quarters. The first meal on board we all used knives and forks, but thereafter they were only supplied to me, while the Japanese fell back upon their chop-sticks. It was a never-failing source of interest to watch their skill in eating under the most difficult circumstances. One morning when the boat was dancing about even more than usual, I came into breakfast to find the steward bringing in some rather underdone fried eggs, and thought that at last I would see the ship's officers stumped in the use of their chop-sticks. Not a bit of it; they had disposed of the eggs in the most unsurpassed manner and were off to their duties before I myself had finished eating.

We left Alexandria with an escort of aero planes to see us safely started, while an observation balloon made fast to a cruiser accompanied us on the first part of our journey. The precautions were not in vain, for two submarines were sighted a short time after we cleared the harbor. The traditional Japanese efficiency was well borne out by the speed with which our crew prepared for action. Every member was in his appointed place and the guns were

---

49   The Umi was part of the 2nd Special Squadron of the Imperial Japanese Navy, assigned to assist the British in the Mediterranean. They engaged submarines, escorted troop ships, and assisted in rescuing survivors when the SS *Transylvania* was sunk.

stripped for action in an incredibly short time after the warning signal. It was when we were nearing the shores of Italy that I had best opportunity to see the destroyers at work. We sighted a submarine which let fly at one of the troopers—the torpedo passing its bow and barely missing the boat beyond it. Quick as a flash the Japanese were after it—swerving in and out like terriers chasing a rat, and letting drive as long as it was visible. We cast around for the better part of an hour, dropping overboard depth charges which shook the little craft as the explosion sent great funnels of water aloft. The familiar harbor of Taranto was a welcome sight when we at length herded our charges in through the narrow entrance and swung alongside the wharf where the destroyers were to take in a supply of fuel preparatory to starting out again on their interminable and arduous task.

Japanese Destroyers of the 2nd Special Squadron entering port at Marseilles

10 Kaba class.

UME (foreground) KUSUNOKI (behind).   1919 *Photo, Comm. Holberton, R.N.*

10 *Kaba* class:—***Kaba*** (Yokosuka), ***Kaede*** (Maidzuru), ***Kashiwa*** and ***Matsu*** (Nagasaki), ***Katsura*** (Kure), ***Kiri*** (Uraga), ***Kusunoki*** and ***Ume*** (Kobe), ***Sakaki*** (Sasebo), and ***Sugi*** (Osaka). All launched February–March, 1915. 665 tons. Dimensions : 260 (*p.p.*) 274 (*o.a.*) × 24 × 7·9 feet. Armament* : 1—4·7 inch, 4—12 pdr., (2 anti-aircraft model) and 4—18 inch tubes. Machinery : 3 sets, 4-cylinder triple expansion and 4 Kansei boilers. Designed H.P. 9,500=30 kts. Fuel : 90 tons coal + 135 tons oil. Complement, 92. These boats are said to have been built in seven months. Built under 1914 Naval Programme, and majority served in Mediterranean during war. 12 Replicas built in Japanese Yards for French Navy—see French *Tribal* class, *Algerien, Arabe,* &c.

*Unofficially reported that proposals have been considered for re-arming these boats (and *Sakura* class) with 2 or 3— 4.7 inch, 45 cal. and 1 AA. gun, to make armament standard with that of recent 2nd Class boats. Not known if any boats have so been modified.

Japanese destroyers passing through the gut at Taranto

CHAPTER NINE

WITH THE FIRST DIVISION IN FRANCE AND GERMANY

I

My transfer to the American army appointed me as captain of field artillery instead of infantry, as I had wished. Just how the mistake occurred I never determined, but once in the field artillery I found that to shift back would take an uncertain length of time, and that even after it was effected I would be obliged to take a course at some school before going up to the line. It therefore seemed advisable to go immediately, as instructed, to the artillery school at Saumur. The management was half French and half American. Colonel MacDonald and Colonel Cross were the Americans in charge, and the high reputation of the school bore testimony to their efficiency. It was the intention of headquarters gradually to replace all the French instructors with Americans, but when I was there the former predominated. It was of course necessary to wait until our officers had learned by actual experience the use of the French guns with which our army was supplied. When men are being taught what to do in combat conditions they apply themselves more attentively and absorb far more when they feel that the officer teaching them has had to test, under enemy fire, the theories he is expounding. The school was for both officers and candidates. The latter were generally chosen from among the non-commissioned officers serving at the front; I afterward sent men down from my battery. The first part of the course was difficult for those who had either never had much mathematical training or had had it so long ago that they were hopelessly out of practice. A number of excellent sergeants and corporals did not have the necessary grounding to enable them to pass the examinations. They should never have been sent, for it merely put them in an awkward and humiliating position—although no stigma could possibly be attached to them for having failed.

The French officer commanding the field work was Major de Caraman. His long and distinguished service in the front lines, combined with his initiative and ever-ready tact, made him an invaluable agent in welding the ideas and methods of France and America. His house was always filled with Americans, and how much his hospitality meant to those whose ties were across the ocean must have been experienced to be appreciated. The homes of France were ever thrown open to us, and the sincere and simple good-will with which we were received has put us under a lasting debt which we should be only too glad to cherish and acknowledge.

Saumur is a delightful old town in the heart of the château country. The river Loire runs through it, and along the banks are the caves in some of which have been found the paintings made by prehistoric man picturing the beasts with which he struggled for supremacy in the dim dark ages. The same caves are many of them inhabited, and their owners may well look with scorn upon the châteaux and baronial castles of whose antiquity it is customary to boast. There is an impressive castle built on a hill dominating the town, and in one of the churches is hung an array of tapestries of unsurpassed color and design. The country round about invited rambling, and the excellent roads made it easy; particularly delightful were the strolls along the river-banks, where patient fisherfolk of every sex and age sat unperturbed by the fact that they never seemed to catch anything. One old lady with a sunbonnet was always to be seen seated on a three-legged stool in the same corner amid the rocks. She had a rusty black umbrella which she would open when the rays of the sun became too searching.

The buildings which were provided for the artillery course had formerly been used by the cavalry school, probably the best known in the world. Before the war army officers of every important nation in the Occident and Orient were sent by their governments to follow the course and learn the method of instruction. My old friend Fitzhugh Lee was one of those sent by the United States, and I found his record as a horseman still alive and fresh in the memory of many of the townspeople.

Soon after the termination of my period of instruction I was in command of C Battery of the Seventh Field Artillery in the Argonne fighting. I was standing one morning in the desolate, shell-ridden town of Landres et St. George watching a column of "dough-boys" coming up the road; at their head limped a battered Dodge car, and as it neared me I recognized my elder brother Ted, sitting on the back seat in deep discussion with his adjutant. I had believed him to be safely at the staff school in Langres recuperating from a wound, but he had been offered the chance to come up in command of his old regiment, the Twenty-sixth, and although registered as only "good for light duty in the service of supply," he had made his way back to the division. While we were talking another car came up and out from it jumped my brother-in-law, Colonel Richard Derby—at that time division surgeon of the Second Division. We were the only three members of the family left in active service since my brother Quentin, the aviator, was brought down over the enemy lines, and Archie, severely wounded in leg and arm, had been evacuated to the United States. I well remember how once when Colonel Derby introduced me to General Lejeune, who was commanding his division, the general, instead of making some remark about my father, said: "I shall always be glad to meet a relative of a man with Colonel Derby's record."

On the 11th of November we had just returned to our original sector after attacking Sedan. None of us placed much confidence in an armistice being signed. We felt that the German would never accept the terms, but were confident that by late spring or early summer we would be able to bring about an unconditional surrender. When the firing ceased and the news came through that the enemy had capitulated, there was no great show of excitement. We were all too weary to be much stirred by anything that could occur. For the past two weeks we had been switched hither and yon, with little sleep and less food, and a constant decrease in our personnel and horses that was never entirely made good but grew steadily more serious. The only bursts of enthusiasm that I heard were occasioned by the automobile trucks and staff cars passing by after dark with their headlights blazing. The joyous shouts of "Lights out!" testified that the reign of darkness was over. Soon the men began building fires and gathering about them, calling "Lights out!" as each new blaze started—a joke which seemed a never-failing source of amusement.

We heard that we were to march into Germany in the wake of the evacuating army and occupy one of the bridge-heads. All this came through in vague and unconfirmed form, but in a few days we were hauled back out of the line to a desolate mass of ruins which had once been the village of Bantheville. We were told that we would have five days here, during which we would be reoutfitted in every particular. Our horses were in fearful shape—constant work in the rain and mud with very meagre allowance of fodder had worn down the toughest old campaigners among them. During the weary, endless night march on Sedan I often saw two horses leaning against each other in utter exhaustion—as if it were by that means alone that they kept on their feet. We were told to indent for everything that we needed to make our batteries complete as prescribed in the organization charts, but we followed instructions without any very blind faith in results—nor did our lack of trust prove unwarranted, for we got practically nothing for which we had applied.

There were some colored troops near by engaged in repairing the roads, and a number of us determined to get up a quartet to sing for the men. We went to where the negroes had built themselves shelters from corrugated-iron sheets and miscellaneous bits of wreckage from the town. We collected three quarters of our quartet and were directed to the mess-shack for the fourth. As we approached I could hear sounds of altercation and a voice that we placed immediately as that of our quarry arose in indignant warning: "If yo' doan' leggo that mess-kit I'll lay a barrage down on yo'!" A platform was improvised near a blazing fire of pine boards and we had some excellent clogging and singing. The big basso had evidently a strong feeling for his steel helmet, and it un-doubtedly added to his picturesqueness—setting off his features with his teeth and eyes gleaming in the firelight.

On the evening of the second day orders came to move off on the following morning. We were obliged to discard much material, for although the two days' rest and food had distinctly helped out the horse situation, we had many animals that could barely drag themselves along, much less a loaded caisson, and our instructions were to on no account salvage ammunition. We could spare but one horse for riding—my little mare—and she was no use for pulling. She was a wise little animal with excellent gaits and great endurance. We were forced to leave behind another mare that I had ridden a good deal on reconnaissances, and that used to amuse me by her unalterable determination to stick to cover. It was almost impossible to get her to cut across a field; she preferred to skirt the woods and had no intention of exposing herself on any sky-line. In spite of her caution it was on account of wounds that she had eventually to be abandoned. I trust that the salvage parties found her and that she is now reaping the reward of her foresight.

We were a sorry-looking outfit as we marched away from Bantheville. My lieutenants had lost their bedding-rolls and extra clothes long since—as every one did, for it was impossible to keep your belongings with you—and although authorized dumps were provided and we were told that anything left behind would be cared for, we would be moved to another sector without a chance to collect our excess and practically everything would have disappeared by the time the opportunity came to visit the cache. But although the horses and accoutrements were in bad shape, the men were fit for any task, and more than ready to take on whatever situation might arise.

Our destination was Malancourt, no great distance away, but the roads were so jammed with traffic that it was long after dark before we reached the bleak, wind-swept hillside that had been allotted to us. It was bitterly cold and we groped about among the shattered barbed-wire entanglements searching for wood to light a fire. There was no difficulty in finding shell-craters in which to sleep—the ground was so pockmarked with them that it seemed impossible that it could have been done by human agency.

This country had been an "active" area during practically all the war, and the towns had been battered and beaten down first by the Boche and then by the French, and lately we ourselves had taken a hand in the further demolition of the ruins. Many a village was recognizable from the encompassing waste only by the sign-board stuck in a mound announcing its name. The next day's march took us through Esne, Montzeville, and Bethainville, and on down to the Verdun-Paris highway. We passed by historic "Dead Man's Hill,"[50] and not far from there we saw the mute reminders of an attack that brought the whole scene vividly back. There were nine or ten tanks, of types varying from the little Renault to the powerful battleship sort. All had been halted by direct hits, some while still far from their objective, others after they had reached the wire entanglements, and there was one that was already astride of the first-line trench. The continual sight of ruined towns and desolated countryside becomes very oppressive, and it was a relief when we began to pass through villages in which many of the houses were still left standing; it seemed like coming into a new world.

At ten in the evening I got the battery into Balaicourt. A strong wind was blowing and the cold was intense, so I set off to try to find billets for the men where they could be at least partly sheltered. The town was all but deserted by its inhabitants, and we managed to provide every one with some degree of cover. Getting back into billets is particularly welcome in very cold or rainy weather, and we all were glad to be held over a day on the wholly mythical plea of refitting. Although the time would not be sufficient to make any appreciable effort in the way of cleaning harness or *materiel*, the men could at any rate heat water to wash their clothes and themselves.

The next day's march we regarded as our first in the advance into Germany to which we had so long looked forward. We found the great Verdun highway which had played such an important part in the defense that broke the back of the Hun to be in excellent shape and a pleasant change from the shell-pitted roads to which we had become accustomed. It was not without a thrill that I rode, at the head of my battery, through the massive south gate of Verdun, and followed the winding streets of the old city through to the opposite portal. Before we had gone many miles the road

---

50   A hill near Verdun, known in French as Le Mort Homme, and German as Toter Mann. A key vantage point and site of much fighting and death.

crossed a portion of the far-famed Hindenburg line which had here remained intact until evacuated by the Boche a few days previously under the terms of the armistice.

We made a short halt where a negro engineer regiment was at work making the road passable. A most hospitable officer strolled up and asked if I wanted anything to eat, which when you are in the army may be classified with Goldberg's "foolish questions."[51] A sturdy coal-black cook brought me soup and roast beef and coffee, and never have I appreciated the culinary arts of the finest French chef as I did that meal, for the food had been cooked, not merely thrown into one of the tureens of a rolling kitchen, which was as much as we had recently been able to hope for.

The negro cook looked as if he would have been able to emulate his French confrére of whom Major de Caraman told me. The Frenchman was on his way to an outpost with a steaming caldron of soup. He must have lost the way, for he unexpectedly found himself confronted by a German who ordered him to surrender. For reply the cook slammed the soup-dish over his adversary's head and marched him back a prisoner. His prowess was rewarded with a Croix de Guerre.

It was interesting to see the German system of defense when it was still intact and had not been shattered by our artillery preparation as it was when taken in an attack. The wire entanglements were miles in depth, and the great trees by the roadside were mined. This was done by cutting a groove three or four inches broad and of an equal depth and filling it with packages of explosive. I suppose the purpose was to block the road in case of retreat. Only a few of the mines had been set off.

Passing through several towns that no longer existed we came to Etain, where many buildings were still standing though completely gutted. The cellars had been converted into dugouts with passages and ramifications added. We were billeted in some German huts on the outskirts. They were well dug in and comfortably fitted out, so we were ready to stay over a few days, as we had been <u>told we should,</u> but at midnight orders were sent round to be pre-

51   Rube Goldberg's comic strip "Foolish Questions" was a very popular comic strip that started in 1909.

pared to march out early.

The country was lovely and gave little sign of the Boche occupation except that it was totally deserted and when we passed through villages all the signs were in German. There was but little originality displayed in naming the streets—you could be sure that you would find a Hindenburg Strasse and a Kronprinz Strasse, and there was usually one called after the Kaiser. The mile-posts at the crossroads had been mostly replaced, but occasionally we found battered metal plaques of the Automobile Touring Club of France. Ever since we left Verdun we had been meeting bands of released prisoners, Italians and Russians chiefly, with a few French and English mingled. They were worn and underfed—their clothes were in rags. A few had combined and were pulling their scanty belongings on little cars, such as children make out of soap-boxes. The motor trucks returning to our base after bringing up the rations would take back as many as they could carry.

We came across scarcely any civilians until we reached Bouligny, a once busy and prosperous manufacturing town. A few of the inhabitants had been allowed to remain throughout the enemy occupation and small groups of those that had been removed were by now trickling in. The invader had destroyed property in the most ruthless manner, and the buildings were gutted. The domestic habits of the Hun were always to me inexplicable—he evidently preferred to live in the midst of his own filth, and many times have I seen recently captured chateaux that had been converted into veritable pigsties.

The inhabitants went wild at our entry—in the little villages they came out carrying wreaths and threw confetti and flowers as they shouted the "Marseillaise." The infantry, marching in advance, bore the brunt of the celebrations. What interested me most were the bands of small children, many of them certainly not over five, dancing along the streets singing their national anthem. It must have been taught them in secret. In the midst of a band were often an American soldier or two, in full swing, thoroughly enjoying themselves. The enthusiasm was all of it natural and uninspired by alcohol, for the Germans had taken with them everything to drink that they had been unable to finish.

Bouligny is not an attractive place—few manufacturing towns are—but we got the men well billeted under water-tight roofs, and we were able to heat water for washing. My striker found a large caldron and I luxuriated in a steaming bath, the first in over a month, and, what was more, I had some clean clothes to pull on when I got out.

One evening, when returning from a near-by village, I met a frock-coated civilian who in-quired of me in German the way to Etain. I asked him who he was and what he wanted. He answered that he was a German but was tired of his country and wished to go almost anywhere else. He seemed altogether too apparent to be a spy, and even if he were I could not make out any object that he could gain. I have often wondered what became of him.

The Boches had evidently not expected to give up their con-quests, for they had built an enormous stone-and-brick fountain in the centre of the town, and chiselled its name, "Hindenburg Brun-nen." Above the German canteen or commissary shop was a great wooden board with "Gott strafe England"—a curious proof of how bitterly the Huns hated Great Britain, for there were no British troops in the sectors in front of this part of the invaded territory.

We worked hard "policing up" ourselves and our equipment during the few days we stayed at Bouligny. One morning all the townsfolk turned out in their best clothes, which had been buried in the cellars or hidden behind the rafters in the attics, to greet the President and Madame Poincare, who were visiting the most important of the liberated towns. It was good to hear the cheering and watch the beaming faces.

On November 21 we resumed our march. Close to the border we came upon a large German cemetery, artistically laid out, with a group of massive statuary in the centre. There were some hero-ic-size granite statues of Boche soldiers in full kit with helmet and all, that were particularly fine. As we passed the stones marking the boundary-line between France and Lorraine there was a tangible feeling of making history, and it was not without a thrill that we entered Aumetz and heard the old people greet us in French while the children could speak only German. The town was gay with the colors of France—produced from goodness knows where. Children were balancing themselves on the barrels of abandoned German cannon and climbing about the huge camouflaged trucks. We were now where France, Luxemburg, and Lorraine meet, and all day we skirted the borders of first one and then the other, halting for the night at the French town of Villerupt. The people went wild when we rode in—we were the first soldiers of the Allies they had seen, for the Germans entered immediately after the declaration of war, and the only poilus the townsfolk saw were those that were brought in as prisoners. We were welcomed in the town hall— the German champagne was abominable but the reception was whole-hearted and the speeches were sincere in their jubilation.

I was billeted with the mayor, Monsieur Georges. After dinner he produced two grimy bottles of Pol Roger—he said that he had been forced to change their hiding-place four times, and had just dug them up in his cellar. They were destined for the night of liberation. Monsieur Georges was thin and worn; he had spent two years in prison in solitary confinement for having given a French prisoner some bread. His eighteen-year-old daughter was imprisoned for a year because she had not informed the authorities as to what her father had done. No one in the family would learn a single word of German. They said that all French civilians were forced to salute the Germans, and each Sunday every one was compelled to appear in the market-place for general muster. The description of the departure of their hated oppressors was vivid—the men behind the lines knew the full portent of events and were sullen and crest-fallen, but the soldiers fresh from the front believed that Germany had won and was dictating her own terms; they came through with wreaths hung on their bayonets singing songs of victory.

I had often wondered how justly the food supplies sent by America for the inhabitants of the invaded districts were distributed. Monsieur Georges assured me that the Germans were scrupulously careful in this matter, because they feared that if they were not, the supplies would no longer be sent, and this would of course encroach upon their own resources, for even the Hun could not utterly starve to death the captured French civilians. The mayor told me of the joy the shipments brought and how when the people went to draw their rations they called it "going to America." We sat talking until far into the night before I retired to the luxury of a real bed with clean linen sheets. There was no trouble whatever in billeting the men—the townsmen were quarrelling as to who should have them.

Next morning, with great regret at so soon leaving our willing hosts, we marched off into the little Duchy of Luxemburg. We passed through the thriving city of Esch with its great iron-mines. The streets were gay with flags, there were almost as many Italian as French, for there is a large Italian colony, the members of which are employed in mining and smelting. Brass bands paraded in our honor, and we were later met by them in many of the smaller towns. The shops seemed well filled, but the prices were very high. The Germans seemed to have left the Luxemburgers very much to themselves, and I have little doubt but that they would have been at least as pleased to welcome victorious Bodies had affairs taken a different turn. Still they were glad to see us, for it meant the end of the isolation in which they had been living and the eventual advent of foodstuffs.

As we rode along, the countryside was lovely and the smiling fields and hillsides made "excursions and alarums"[52] seem remote indeed. It felt unnatural to pass through a village with unscarred church spires and houses all intact—such a change from battered, glorious France.

We were immediately in the wake of the German army, and taken by and large they must have been retiring in good order, for they left little behind. Our first night we spent at the village of Syren, eight kilometres from the capital of the Duchy. Billeting was not so easy now, for we were ordered to treat the inhabitants as neutrals, and when they objected we couldn't handle the situation as we did later on in Germany. No one likes to have soldiers or civilians quartered on him, and the Luxemburgers were friendly to us only as a matter of policy. Fortunately, the chalk marks of the Boche billeting officers had not been washed off the doors, and these told us how many men had been lodged in a given house.

52   Stage directions from Shakespeare, in *Henry V.*

In my lodging I was accorded a most friendly reception, for my hostess was French. Her nephew had come up from Paris to visit her a few months before the outbreak of the war, and had been unable to get back to France. To avoid the dreaded internment camp he had successfully passed as a Luxemburger. In the regiment there were a number of men whose parents came from the Duchy; these and a few more who spoke German acquired a sudden popularity among their comrades. They would make friends with some of the villagers and arrange to turn over their rations so that they would be cooked by the housewife and eaten with the luxurious accompaniment of chair and table. The diplomat would invite a few friends to enjoy with him the welcome change from the "slum" ladled out of the cauldrons of the battery rolling kitchen. I had always supposed that I had in my battery a large number of men who could speak German—a glance over the pay-roll would certainly leave that impression—but when I came to test it out, I found that I had but four men who spoke sufficiently well to be of any use as interpreters.

Next morning we made a winding, roundabout march to Trintange. Here we were instructed to settle down for a week or ten days' halt, and many worse places might have been chosen. The country was very broken, with hills and ravines. Little patches of woodland and streams dashing down rocky channels on their way to join the Moselle reminded one of Rock Creek Park in Washington. The weather couldn't be bettered; sharp and cold in the early morning with a heavy hoarfrost spreading its white mantle over everything, then out would come the sun, and the hills would be shrouded in mist.

My billeting officer had arranged matters well, so we were comfortably installed and in good shape to "police up" for the final leg of the march to Coblenz. I had now my full allowance of officers—Lieutenants Furness, Brown, Middleditch, and Pearce. In active warfare discipline while stricter in some ways is more lax in others, and there were many small points that required furbishing. Close order drill on foot is always a great help in stiffening up the men, and such essentials as instruction in driving and in fitting harness required much attention. In the American army much less responsibility is given to the sergeants and corporals than in the British, but even so the spirit and efficiency of an organization must depend largely on its noncommissioned officers. We were fortunate in having an unusually fine lot—Sergeant Cushing was a veteran of the Spanish War. He had been a sailor for many years, and after he left the sea he became chief game warden of Massachusetts. In time of stress he was a tower of strength and could be counted upon to set his men an example of cool and judicious daring. The first sergeant, Armstrong, was an old regular army man, and his knowledge of drill and routine was invaluable to us. He thoroughly understood his profession, and was remarkably successful in training raw men. Sergeants Grumbling, Kubelis, and Bauer were all of them excellent men, and could be relied upon to perform their duty with conscientious thoroughness under the most trying conditions.

One afternoon I went in to Luxemburg with Colonel Collins, the battalion commander. The town looks thoroughly mediaeval as you approach. It might well have been over its castle wall that Kingsley's knight spurred his horse on his last leap; as a matter of fact the village of Altenahr, where the poet laid the scene, is not so many miles away. The town is built along the ragged cliffs lining a deep, rocky canyon spanned by old stone bridges. The massive entrance-gates open upon passages tunnelled through the hills, and although the modern part of the town boasts broad streets and squares, there are many narrow passageways winding around the ancient quarter.

I went into a large bookstore to replenish my library, and was struck by the supply of post-cards of Marshal Foch and Kitchener and the King and Queen of Belgium. All had been printed in Leipzig, and when I asked the bookseller how that could be, he replied that he got them from the German commercial travellers. He said that he had himself been surprised at the samples shown him, but the salesman had remarked that he thought such post-cards would have a good sale in Luxemburg, and if such were the case "business was business," and he was prepared to supply them. There was even one of King Albert standing with drawn sword, saying: "You shall not violate the sacred soil of my country." A publication that also interested me was a weekly paper brought out in Hamburg and written in English. It was filled with jokes, beneath which were German notes explaining any difficult or idiomatic words and phrases. With all their hatred of England the Huns still continued to learn English.

Thanksgiving Day came along, and we set to work to provide some sort of a special feast for the men. It was most difficult to do so, for the exchange had not as yet been regulated and the lowest rate at which we could get marks was at a franc, and usually it was a franc and a quarter. Some one opportunely arrived from Paris with a few hundred marks that he had bought at sixty centimes. For the officers we got a suckling pig, which Mess Sergeant Braun roasted in the priest's oven. He even put the traditional baked apple in its mouth, a necessary adjunct, the purpose of which I have never discovered, and such stuffing as he made has never been equalled. We washed it down with excellent Moselle wine, for we were but a couple of miles from the vineyards along the river. In the afternoon I borrowed a bicycle from the burgomaster and trailed over to Elmen, where I found my brother just about to sit down to his Thanksgiving dinner served up by two faithful Chinamen, who had come to his regiment in a draft from the West Coast.

After doing full justice to his fare I wended my way back to Trintange in the rain and dark.

The next day we paid the men. For some it was the first time in ten months. To draw pay it was necessary to sign the pay-roll at the end of one month and be on hand at the end of the following month to receive the money. No one could sign unless his service record was at hand, and as this was forwarded to the hospital "through military channels" when a man was evacuated sick or wounded, it rarely reached his unit until several months after he returned. It may easily be seen why it was that an enlisted man often went for months without being able to draw his pay. This meant not only a hardship to him while he was without money, but it also followed that when he got it he had a greater amount than he could possibly need, and was more than apt to gamble or drink away his sudden accession of wealth. We always tried to make a man who had drawn a lot of back pay deposit it or send it home. Mr. Harlow, the Y.M.C.A. secretary attached to the regiment, helped us a great deal in getting the money transferred to the United States. The men, unless they could spend their earnings immediately, would start a game of craps and in a few days all the available cash would have found its way into the pocket of the luckiest man. They would throw for appallingly high stakes. On this particular pay-day we knew that the supply of wine and beer in the village was not sufficient to cause any serious trouble, and orders were given that no cognac or hard liquor should be sold. A few always managed to get it—all precautions to the contrary notwithstanding.

II

On the 1st of December we once more resumed our march and at Wormeldange crossed over the Moselle River into Hunland. The streets of the first town through which we passed were lined with civilians, many of them only just out of uniform, and they scowled at us as we rode by, muttering below their breath. A short way out and we began to meet men still in the field-gray uniform; they smiled and tried to make advances but our men paid no attention. When we reached Onsdorf, which was our destination, the billeting officer reported that he had met with no difficulty.

The inhabitants were most effusive and anxious to please in every way. Of course they were not Prussians, and no doubt were heartily tired and sick of war, but here, as throughout, their attitude was most distasteful to us—it was so totally lacking in dignity. We could not tell how much they were acting on their own initiative and to what extent they were following instructions. Probably there was something of both back of their conduct. Warnings had been issued that the Germans were reported to be planning a wholesale poisoning of American officers, but I never saw anything to substantiate the belief.

Next morning we struck across to the Saar River and followed it down to its junction with the Moselle. The woods and ravines were lovely, but from the practical standpoint the going was very hard upon the horses. We marched down through Treves, the oldest town in Germany, with a population of about thirty thousand.

In the fourth century of our era Ausonius referred to it as "Rome beyond the Alps," and the extent and variety of the Roman remains would seem to justify the epithet.[53] We were halted for some time beside the most remarkable of these, the Porta Nigra, a huge fortified gateway, dating from the first century A. D. The cathedral is an impressive conglomeration of the architecture of many different centuries—the oldest portion being a part of a Roman basilica of the fourth century, while the latest additions of any magnitude were made in the thirteenth. Most famous among its treasures is the "holy coat of Treves," believed by the devout to be the seamless garment worn by Christ at the crucifixion.[54] The predominant religion of the neighborhood is the Roman Catholic, and on the occasions when the coat is exhibited the town is thronged by countless pilgrims.

Leaving Treves we continued down along the river-bank to Rawen Kaulin, where we turned inland for a few miles and I was assigned to a village known as Eitelsbach. The inhabitants were badly frightened when we rode in—most of the men hid and the women stood on the door-steps weeping. I suppose they expected to be treated in the manner that they had behaved to the French and Belgians, and as they would have done by us had the situation been reversed. When they found they were not to be oppressed they became servile and fawning. I had my officers' mess in the schoolmaster's house. He had been a non-commissioned officer of infantry, and yet he wanted to send his daughters in to play the piano for us after dinner. We would have despised the German less if he had been able to "hate" a little more after he was beaten and not so bitterly while he felt he was winning.

The country through which we marched during the next few days was most beautiful. We followed the winding course of the river, making many a double "S" turn. The steep hills came right to the bank; frequently the road was cut into their sides. A village was tucked in wherever a bit of level plain between the foot of the hill

53   Decimius Magnus Ausonius (310-395) was a Roman poet and rhetorician. He tutored Emperor Gratian.
54   Not to be confused with the famous Shroud of Turin. This was a relic reputed to have been the "seamless garment" worn by Christ at the resurrection.

and the river permitted. When the slopes gave a southern exposure they were covered with grape-vines, planted with the utmost precision and regularity. Every corner and cranny among the rocks was utilized. The original planting must have been difficult, for the soil was covered with slabs of shale. The cultivator should develop excellent lungs in scaling those hillsides. The leaves had fallen and the bare vines varied in hue from sepia brown to wine color, with occasional patches of evergreen to set off the whole. Once or twice the road left the river to cut across over the mountains, and it cost our horses much exertion to drag the limbers up the steep, slippery trail. It was curious to notice the difference between those who dwelt along the bank and the inhabitants of the upland plateau. The latter appeared distinctly more "outlandish" and less sleek and prosperous. The highlands we found veiled in mist, and as I looked back at the dim outlines of horse and man and caisson, it seemed as if I were leading a ghost battery.

We were in the heart of the wine country, and to any one who had enjoyed a good bottle of Moselle such names as Berncastel and Piesport had long been familiar. In the former town I was amused on passing by a large millinery store to see the proprietor's name was Jacob Astor. The little villages inevitably recalled the fairy-tales of Hans Andersen and the Grimm brothers. The raftered houses had timbered balconies that all but met across the crooked, winding streets through which we clattered over the cobblestones. Capping many of the beams were gargoyles, demons, and dwarfs, and a galaxy of strange creatures were carved on the ends of the gables that jutted out every which way. The houses often had the date they were built and the initials of the couple that built them over the front door, frequently with some device. I saw no dates that went further back than the late sixteen hundreds, though many of the houses doubtless were built before then. The doors in some cases were beautifully carved and weathered. The old pumps and wells, the stone bridges, and the little wayside shrines took one back through the centuries. To judge by the records carved on wall and house, high floods are no very uncommon occurrence—the highest I noticed was in 1685, while the last one of importance was credited to 1892.

We were much surprised at the well-fed appearance of the population, both old and young, for we had heard so much of food shortages, and the Germans when they surrendered had laid such stress upon it. As far as we could judge, food was more plentiful than in France. Rubber and leather were very scarce, many of the women wore army boots, and the shoes displayed in shop-windows appeared made of some composition resembling pasteboard. The coffee was evidently ground from the berry of some native bush, and its taste in no way resembled the real. Cigars were camouflaged cabbage-leaves, with little or no flavor, and the beer sadly fallen off from its pre-war glory. Still, in all the essentials of life the inhabitants appeared to be making out far better than we had been given to believe.

We met with very little trouble. There were a few instances where people tried to stand out against having men billeted in their houses, but we of course paid no attention except that we saw to it that they got more men than they would have under ordinary circumstances. Every now and then we would have amusing side-lights upon the war news on which the more ignorant Boches had been fed. A man upon whom several of my sergeants were quartered asked them if the Zeppelins had done much damage to New York; and whether Boston and Philadelphia had yet been evacuated by the Germans—he had heard that both cities had been taken and that Washington was threatened and its fall imminent.

Our men behaved exceedingly well. Of course there were individual cases of drunkenness, but very few considering that we were in a country where the wine was cheap and schnapps plentiful. There were the inevitable A.W.O.L.'s and a number of minor offenses, but I found that by making the prisoner's life very unattractive—seeing to it that they performed distasteful "fatigues," giving them heavy packs to carry when we marched, and allowing them nothing that could be construed as a delicacy—I soon reformed the few men that were chronically shiftless or untidy or late. When not in cantonments the trouble with putting men under arrest is that too often it only means that they lead an easier life than their comrades, and it takes some ingenuity to correct this situation. Whenever it was in any way possible an offender was dealt with in the battery and I never let it go further, for I found it made for much better spirit in a unit.

The men were a fine lot, and such thoroughgoing Americans, no matter from what country their parents had come. One of my buglers had landed in the United States only in 1913; he had been born and brought up on the confines of Germany and Austria, and yet when a large German of whom he was asking the way said, "You speak the language well—your parents must be German," the unhesitating reply was: "Well, my mother was of German descent!" The battery call read like a League of Nations, but no one could have found any cause of complaint in lack of loyalty to the United States.

The twelfth day after we had crossed over the river from Luxemburg found us marching into Coblenz. We were quartered in large brick barracks in the outskirts of the city. The departing Germans had left them in very bad shape, and Hercules would have felt that cleaning the Augean stables was a light task in comparison. However, we set to work without delay and soon had both men and horses well housed. Life in the town was following its normal course; the stores were well stocked and seemed to be doing a thriving trade. We went into a cafe where a good orchestra was playing and had some very mediocre war beer, and then I set off in search of the Turkish bath of which I was much in need. The one I found was in charge of an ex-submarine sailor, and when I was shut in the steam-room I wondered if he were going to try any "frightfulness," for I was the only person in the bath. My last one had been in a wine-vat a full week before, and I was ready to risk anything for the luxury of a good soak.

Orders to march usually reached us at midnight—why, I do not know; but we would turn in with the belief that we would not move on the following day, and the next we knew an orderly from regimental headquarters would wake us with marching instructions, and in no happy frame of mind we would grumblingly tumble out to issue the necessary commands. Coblenz proved no exception to this rule. As we got under way, a fine rain was falling that was not long in permeating everything. Through the misty dripping town the "caissons went rolling along,"[55] and out across the Pfaffendorf bridge, with the dim outlines of the fortress of Ehrenbreitstein towering above us. The men were drowsy and cold. I heard a few disparaging comments on the size of the Rhine. They had heard so much talk about it that they had expected to find it at least as large as the Mississippi. We found the slippery stones of the street ascending from the river most difficult to negotiate, but at length everything was safely up, and we struck off toward the bridge-head position which we were to occupy for we knew not how long. The Huns had torn down the sign-posts at the cross-roads; with what intent I cannot imagine, for the roads were not complicated and were clearly indicated on the maps, and the only purpose that the sign-posts could serve was to satisfy a curiosity too idle to cause us to calculate by map how far we had come or what distance lay still before us. A number of great stone slabs attracted our attention; they had been put up toward the close of the eighteenth century and indicated the distance in hours. I remember one that proclaimed it was three hours to Coblenz and eighteen to Frankfort. I have never seen elsewhere these records of an age when time did not mean money.

The march was in the nature of an anticlimax, for we had thought always of Coblenz as our goal, and the good fortune in which we had played as regarded weather during our march down the valley of the Moselle had made us supercritical concerning such details as a long, wearisome slogging through the mud in clumsy, water-logged clothes. At length we reached the little village of Niederelbert and found that Lieutenant Brown, whose turn it was as billeting officer, had settled us so satisfactorily that in a short time we were all comfortably steaming before stoves, thawing out our cramped joints.

With the exception of Lieutenant Furness my officers belonged to the Reserve Corps, and we none of us looked forward to a long tour of garrison duty on the Rhine or anywhere else. Furness, who

---

55   "The Army Goes Rolling Along" is the official song of the United States Army. It was adapted from the 1908 work "The Caissons Go Rolling Along," written by Edmund Gruber, which was incorporated into John Phillip Sousa's "U.S. Field Artillery March." A caisson is a two-wheeled cart to support the trail of an artillery piece.

had particularly distinguished himself in liaison work with the infantry, held a temporary commission in the regular army, but he was eager to go back to civil life at the earliest opportunity. In Germany the prospect was doubly gloomy, for there would be no intercourse with the natives such as in France had lightened many a weary moment. Several days later regimental headquarters coveted our village and we were moved a few miles off across the hills to Holler. We set to work to make ourselves as snug and comfortable as possible. I had as striker a little fellow of Finnish extraction name Jahoola, an excellent man in every way, who took the best of care of my horse and always managed to fix up my billet far better than the circumstances would seem to permit.

The days that followed presented little variety once the novelty of the occupation had worn off. The men continued to behave in exemplary fashion, and the Boche gave little trouble. As soon as we took up our quarters we made the villagers clean up the streets and yards until they possessed a model town, and thereafter we "policed up" every untidiness of which we might be the cause, and kept the inhabitants up to the mark in what concerned them. The head of the house in which I was lodged in Niederelbert told me that his son had been a captain in the army but had deserted a fortnight before the armistice and reached home in civilian clothes three weeks in advance of the retreating army. Of course he was not an officer before the war—not of the old military school, but the fact that he and his family were proud of it spoke of a weakening discipline and morale.

Now that we had settled down to a routine existence I was doubly glad of such books as I had been able to bring along. Of these, O. Henry was the most popular. The little shilling editions were read until they fell to pieces, and in this he held the same position as in the British army. I had been puzzled at this popularity among the English, for much of his slang must have been worse than Greek to them. I also had *Charles O'Malley* and *Harry Lorrequer*, Dumas' *Dame de Monsereau* and *Monte Cristo*, Flaubert's *Education Sentimentale*, Gibbon's *Rise and Fall*, and Borrow's *Zincali*.[56] These with the Oxford Books of French and English verse

---

56   Charles Lever (1806-1872) was an Irish novelist who wrote *Charles*

and a few Portuguese and Spanish novels comprised my library, a large one considering the circumstances. It was always possible to get books through the mail, although they were generally many months en route.

Soon after we reached the bridge-head, officers of the regular army began turning up from the various schools whither they had been sent as instructors. We all hoped to be released in this manner, for we felt that the garrison duty should be undertaken by the regulars, whose life business it is, in order to allow the men who had left their trades and professions to return to their normal and necessary work. In the meantime we set out to familiarize ourselves with the country and keep our units in such shape that should any unforeseen event arise we would be in a position to meet it. The horses required particular attention, but one felt rewarded on seeing their improvement. There were many cases of mange which we had been hitherto unable to properly isolate, and good fodder in adequate quantity was an innovation.

For the men we had mounted and unmounted drill, and spent much time in getting the accoutrements into condition for inspection. During part of the march up rations had been short, and for a number of days were very problematical. Sufficient boots and clothing were also lacking and we had had to get along as best we could without. Now that we were stationary our wants were supplied, and the worst hardship for the men was the lack of recreation. A reading-room was opened and a piano was procured, but there was really no place to send them on short passes; nothing for them to do on an afternoon off. When I left, trips down the Rhine were being planned, and I am sure they proved beneficial in solving the problem of legitimate relaxation and amusement.

My father had sent my brother and myself some money to use in trying to make Christmas a feast-day for the men. It was difficult to get anything, but the Y. M. C. A. very kindly helped me out in procuring chocolates and cigarettes, and I managed to

O'Malley, The Irish Dragoon is an 1841 novel about an Irishman in the peninsular war. Confessions of Harry Lorequer is a similar 1839 tale about a young British officer. George Borrow (1803-1881) was a linguist, traveler, and Bible salesman who wrote The Zincali: An Account of the Gypsies in Spain in 1843. We will presume the other titles listed are better known to the Reader.

buy a couple of calves and a few semi-delicacies in the local market. While not an Arabian Night feast, we had the most essential adjunct in the good spirits of the men, who had been schooled by their varied and eventful existence of the past eighteen months to make the most of things.

In the middle of January my brother and I left for Paris. I was very sorry to leave the battery, for we had been through much together, but in common with most reserve officers I felt that, now that the fighting was over, there was only one thing to be desired and that was to get back to my wife and children. The train made light of the distance over which it had taken us so long to march, and the familiar sight of the friendly French towns was never more welcome. After several months on duty in France and Italy, I sailed on a transport from Brest, but not for the wonderful home-coming to which I had so long looked forward.

THE END

Christmas Card from British Mesopotamian Expeditonary Force

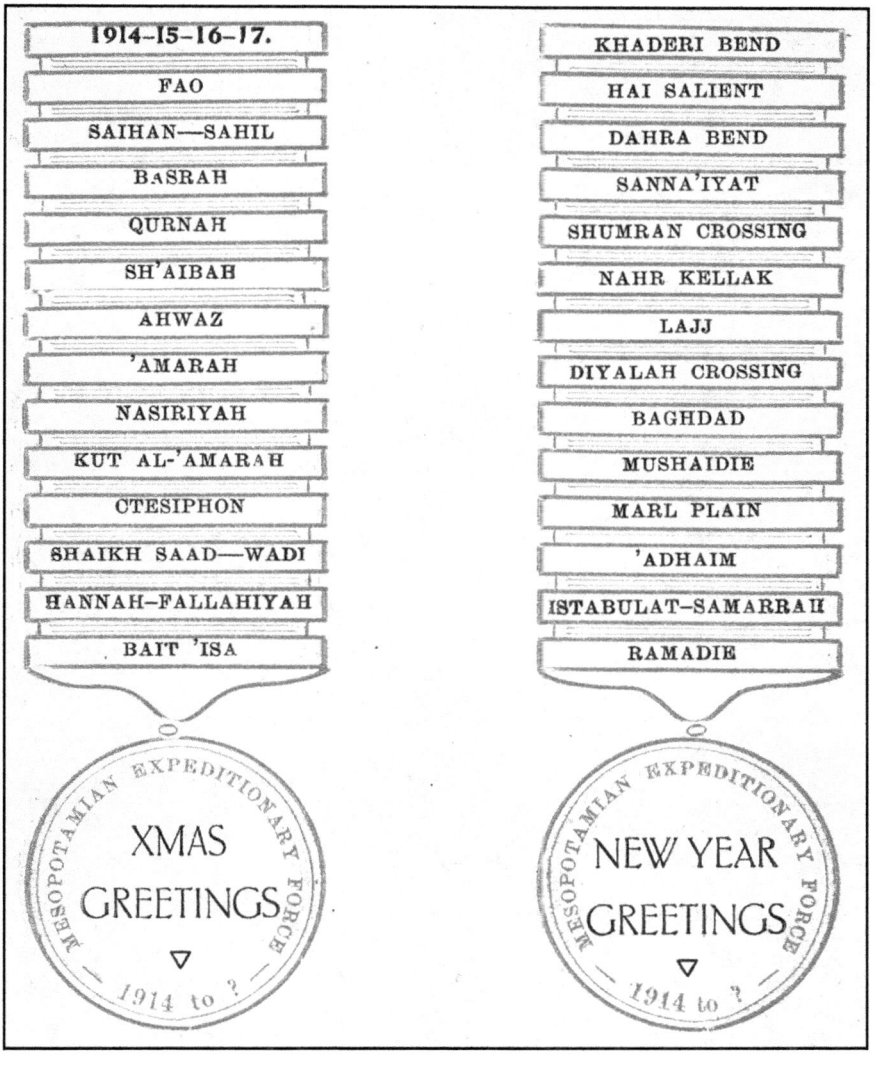

1914–15–16–17.	KHADERI BEND
FAO	HAI SALIENT
SAIHAN—SAHIL	DAHRA BEND
BASRAH	SANNA'IYAT
QURNAH	SHUMRAN CROSSING
SH'AIBAH	NAHR KELLAK
AHWAZ	LAJJ
'AMARAH	DIYALAH CROSSING
NASIRIYAH	BAGHDAD
KUT AL-'AMARAH	MUSHAIDIE
CTESIPHON	MARL PLAIN
SHAIKH SAAD—WADI	'ADHAIM
HANNAH–FALLAHIYAH	ISTABULAT–SAMARRAH
BAIT 'ISA	RAMADIE

MESOPOTAMIAN EXPEDITIONARY FORCE

XMAS GREETINGS
▽
1914 to ?

MESOPOTAMIAN EXPEDITIONARY FORCE

NEW YEAR GREETINGS
▽
1914 to ?

The War In

Explicit iste liber, scriptor sit crimine liber, Christus scriptorem custodiat ac det honorem

Ὥσπερ ξένοι χαίρουσιν ἰδεῖν
πατρίδα, οὕτως καὶ οἱ γράφοντες
ἰδεῖν βιβλίου τέλος

श्रीकृष्णार्पणमस्तु

書成矣，感盡天地

סלוע ארוב לאל חבש סלשנו סת

"of making many books there
is no end; and much study is a
weariness of the flesh"
- Ecclesiastes 12:12

Bulkington Books

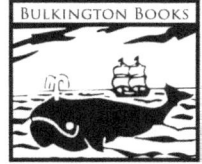